AUTOMATIC AIR QUALITY MONITORING SYSTEMS

AUTOMATIC AIR QUALITY MONITORING SYSTEMS

proceedings of the Conference held at the National Institute of Public Health, Bilthoven, The Netherlands, 5–8 June, 1973

Edited by

T. SCHNEIDER

ELSEVIER SCIENTIFIC PUBLISHING COMPANY

Amsterdam – London – New York

1974

ELSEVIER SCIENTIFIC PUBLISHING COMPANY
Jan van Galenstraat 335
P.O. Box 1270, Amsterdam, The Netherlands

AMERICAN ELSEVIER PUBLISHING COMPANY, INC.
52 Vanderbilt Avenue
New York, New York 10017

Library of Congress Catalog Card Number: 73-90962

ISBN 0-444-41202-6

Printed in The Netherlands

CONTENTS

FOREWORD

The symposium was organized by the National Institute of Public Health (Rijks Instituut voor de Volksgezondheid) in Bilthoven. The co-sponsors were the World Health Organization, Division of Environmental Health, Geneva, the Regional Office Europe, Copenhagen, and the International Reference Center, Research Triangle Park, N.C., U.S.A.

The purpose of the symposium was to exchange knowledge on existing and planned automated air quality monitoring systems and the analysis of the air pollution data. As a follow-up to the symposium, an international working group will be formed to study the use of monitoring systems (planning, construction, operation and data analysis). Membership of this group will be open to National Research Institutes and Government Agencies.

In the order in which the sessions were held, the session chairmen were: D.Shearer, U.S.A.; G.J.Cleary, W.H.O., Geneva; L.J.Brasser, The Netherlands; and H.Stewart, Great Britain.

It is a pleasure to recognize the excellent assistance that was given by the Scientific Secretaries: Y.Buurma, N.D.van Egmond, A.Minderhoud and H.J.W.J.van de Wiel, all staff members of the Laboratory for Air, Water and Soil Research. Thanks are also due to the Organizing Committee, Mrs.L.M.Oostwouder and J.Th.Konter, its Secretariat, Mrs.C.Boom, Mrs.T.Frankhuizen and Mrs.W.A.Wermenbol, and the following co-operators at the Institute: R.van 't Hoenderdaal, Mrs.K.M.Rijksen, J.H.van Stekelenburgh, J.F.Worrell and C.G.van Elst.

October 1973

T.Schneider

PARTICIPANTS

BELGIUM

Henri Bultynck
SCK/CEN Boeretang 200
B-2400 MOL

Gilbert Cocquyt
SCK/CEN Boeretang 200
B-2400 MOL

Jean Kretzschmar
SCK/CEN Boeretang 200
B-2400 MOL

X.de Maere
SCK/CEN Boeretang 200
B-2400 MOL

J.Bouquiaux
Institut d'Hygiène et d'Epidémiodologie
14 Rue Juliette Wytsman
1050 BRUXELLES

J.Exteyl
Ministerie van Tewerkstelling en Arbeid
51/53 Belliardstraat
1040 BRUXELLES

J.Vandervee
SCK/CEN Boeretang 200
B-2400 MOL

CANADA

Henk W.de Koning
Rm.242
Environmental Health Centre
Tunney's Pasture
OTTAWA K1A 0H3, Ontario

D.M.Whelpdale
Atmospheric Environment Service
4905 Dufferin Street
Downsview, Ontario, M3H5T4

CZECHOSLOVAKIA

V.Štěpánek
Czechoslovak Research and Development Centre
for Environmental Pollution Control
UNO Programme
Hybernská 30
PRAGUE 1

FEDERAL REPUBLIC OF GERMANY

D.Jost
Pilotstation am Institut für Meteorologie
Universität Frankfurt
Feldbergstrasse 47
D-6 FRANKFURT

E.Weber
Bundesministerium des Innern (U III 2)
Rheindorferstrasse 198
53 BONN

H.Schwegler
Bayerisches Landesamt für Umweltschutz
Rosenkavelierplatz 3
8000 MUNCHEN 81

FINLAND

M.JANTUNEN
The Institute of Occupational Health
Haartmaninkatu 1
HELSINKI 25

HUNGARY

Iván Polgár
Ministry of Building and Urban Development
Beloiannisz 2-4
BUDAPEST V

JAPAN

Yasumasa Kumazawa
Chief Engineer Environmental Pollution Countermeasures Section
Osaka Prefectural Government
Otemaeno-cho
Higashi-ku
OSAKA

YUGOSLAVIA

Boris Wilder
The Institute of Medical Research and Industrial Hygiëne
158 Mose Pijade
ZAGREB

THE NETHERLANDS

L.J.Brasser
IG/TNO
Postbus 214
DELFT

A.J.H.C.Brüll
Provinciale Waterstaat
Stadhuisstraat 4
MAASTRICHT

F.W.van Dijk
B.V. Philips Nederland
Boschdijk 525
EINDHOVEN

J.J.Wilting
Philips Industries
P.I.T. TQ5
EINDHOVEN

S.Zwerver
Provinciale Waterstaat
Koningskade 1-2
's GRAVENHAGE

P.Spaander
Ruysdaellaan 33
HUIS TER HEIDE

S.Kruizinga
KNMI
Utrechtseweg 297
DE BILT

D.J.Kroon
Philips Research Lab.
EINDHOVEN

J.Spaander
RIV
Postbus 1
BILTHOVEN

H.Cohen
RIV
Postbus 1
BILTHOVEN

T.Schneider
RIV
Postbus 1
BILTHOVEN

IJ.Buurma
RIV
Postbus 1
BILTHOVEN

N.D.van Egmond
RIV
Postbus 1
BILTHOVEN

H.J.van de Wiel
RIV
Postbus 1
BILTHOVEN

P.Gootjes
RIV
Postbus 1
BILTHOVEN

A.Minderhoud
RIV
Postbus 1
BILTHOVEN

P.van Ulden
KNMI
Utrechtseweg 297
DE BILT

NORWAY

Jack Nordø
Norwegian Institute for Air Research
P.B.15
2077 KJELLER

POLAND

Karl Grzybowski
Director Sanitary Epidemiological Station
Katowice
c/o Project Co-Manager
Environmental Pollution Abatement Centre
KATOWICE

SWITZERLAND

A.Deuber
Institut für Hygiene und Arbeitsphysiologie ETH
Clausiusstraat 25
8006 ZÜRICH

UNITED KINGDOM

H.Stewart
Warren Spring Laboratory
Dept. of Trade and Industry
Gunnels Wood Road
STEVENAGE (Hertfordshire)

USA

R.E.Ruff
EPA
Data Manager, RAPS
Meteorology Laboratory
RESEARCH TRIANGLE PARK
North Carolina 27711

Franz Burmann
EPA/NERC-RTP
National Environmental Research Center
RESEARCH TRIANGLE PARK
North Carolina 27711

D.Shearer
EPA
RESEARCH TRIANGLE PARK
North Carolina 27711

COMMISSION OF EUROPEAN COMMUNITIES

A.Berlin
CCE
29 Rue Aldringen
LUXEMBOURG

J.Smeets
29 Rue Aldringen
LUXEMBOURG

WORLD HEALTH ORGANIZATION

Michael J.Suess,
Regional Officer for Environmental Health
8 Scherfigsvej
DK-2100 COPENHAGEN

S.Greenfield
EPA
401 M.Street SW
WASHINGTON DC 20460

G.J.Cleary,
Scientist Environmental Pollution
Division of Environmental Health
WHO
1211 GENEVE 27

A REVIEW OF AUTOMATED MONITORING SYSTEMS FOR AIR QUALITY*

FRANZ J. BURMANN

U.S. Environmental Protection Agency, National Environmental Research Center Research Triangle Park, North Carolina 27711, U.S.A.

SUMMARY

To have a totally automated monitoring system for air quality, four major elements are necessary. The first element consists of site selection and sample intake system to obtain a representative sample. The second component required is a reliable sensor capable of performing within the required specifications. The third element needed is the transmittance of data from the sensor to the central facility and the command function (return from the central facility) to the measuring device. The fourth step is a mechanism by which the obtained data can be validated, i.e., assurance that the data from its origination with the final print-out has not been altered and was obtained by a properly working instrument.

Obviously all of the elements are equally important for a completely automated monitoring system to yield reliable representative measurements of air quality. The weakest and most critical links in the chain, however, are a properly operating instrument and a representative site. Without these two components every monitoring system – manual, semi- or fully-automatic – is deficient.

A contributing factor to this dilemma is that both guidelines for locating an instrument and the selection of any air monitoring instrument depend on the objective for which monitoring is conducted.

Therefore, a plea has to be made to bring the users of the data – meteorologists, air pollution experts, city planners, health departments responsible for designing and operating monitoring systems as soon as possible so that they may interact with each other and assure full and proper utilization of the generated data.

---

* The mention of company or brand names is not to be considered as endorsement or recommendation for use by the Environmental Protection Agency.

Gathering air quality data *is* serious business because the generated data touches everybody's life by affecting not only health and welfare but also economics, esthetics, and the quality of life itself.

INTRODUCTION

Before discussing automated monitoring systems for air quality, it seems only proper to spend a few moments recalling why people of all nations are concerned with clean air. Clean air is, probably, the most necessary component of man's environment. Man can do without food and water for days. He cannot do without air for more than a few minutes. He must breathe, even if, as has occurred in the past, it kills him.

Let us recall that the first reported tragedy of air pollution occurred in 1930 in the Meuse River Valley, Belgium, where 60 persons died. Let us also recall that in 1948 polluted air sickened more than 40 percent of the population of Donora, Pennsylvania, and killed 22; that during 4 days in 1952 an estimated 4000 people in London died prematurely because of air pollution; and that during a 3-day period in 1966 air pollution killed 168 persons in New York City. It took many years, however, before the general public suspected that foul air is a contributing factor in many respiratory diseases.

Even if unclean air does not affect people's health, it often makes their lives miserable or less enjoyable. Air pollution is responsible for offensive odors, produces damage costing billions of dollars each year to property such as metal, paint, and cloth, and is responsible for death and damage of all kinds to vegetation, from petunias to pine and oak trees.

Industrial processes and combustion sources are the main contributors to air pollution. Because one of the biggest single sources of air contaminants is the automobile, the more industrialized countries with densely populated areas and widespread vehicle use experience the greatest problems.

Surveillance of air quality has become an increasingly important function of any air pollution control agency. Initially the agency must determine the nature and magnitude of the area's air pollution problem, gather data to set standards, and determine the extent to which these standards are being exceeded. Following the adoption of regulations to reduce emissions, atmospheric surveillance is required to evaluate the progress towards achieving the ambient air quality goals. An effective atmospheric surveillance program capable of measuring changes and trends in air quality requires reliable ambient air pollution measurements over prolonged periods.

ATMOSPHERIC MONITORING ACTIVITIES

For this discussion, atmospheric monitoring activities are categorized into three areas - research, surveillance, and special purpose monitoring.

Research Monitoring

The first step in any air quality investigation is to identify the pollutant that requires attention. Included in this task are studies to determine the movement of pollutants from source to receptor.

Monitoring systems that achieve these objectives generally are quite sophisticated and may require a highly skilled operator at all times. If the gathered data indicates that a problem exists, the degree and kind of pollutant effect on people, on other living organisms, and on materials has to be established. These studies require an interdisciplinary effort and the coordination of many programs.

Surveillance Monitoring

Ambient air surveillance provides the information necessary to judge the effectiveness of emission controls; to evaluate the progress toward the attainment of air quality goals; and, where air quality standards exist, to determine to what extent the standards have been exceeded. Included in this activity is "trend monitoring". Basically, trend monitoring involves the use of an air surveillance network that continuously gathers ambient air data at a specific location for many years; thus, pollutant trends can be determined for a metropolitan area, region, state, country, or the world.

Special Purpose Monitoring

Among the tasks of special purpose monitoring are mobile monitoring activities, aerial surveys, and responses to emergency episodes, i.e., to provide a warning for all unforeseen high pollution levels so they can be abated before reaching a crisis and to determine the scope of danger because of any type of pollution emergency, such as accidental spills.

Coordination among these three monitoring activities to assure compatibility of data is a difficult problem. For example, when performing research monitoring to ascertain health-related pollutant levels, one should be concerned that the data may also be utilized for some other purpose such as trend analysis. Consideration should also be given to the possibility that, if a standard results from an investigation, a large number of monitoring stations capable of producing comparable results may be required. Therefore, before any automated monitoring system can be applied, the objective for which data are being collected must be clearly defined, the future ramifications must be known and the methodology must be carefully selected.

Many of today's aerometric techniques are based on classical analytical methods, both chemical and physical. Before discussing monitoring methods applicable primarily for a specific pollutant, I will briefly outline operational principles that have found broad application in continuous air monitoring systems.

COMMON PRINCIPLES APPLIED IN CONTINUOUS AIR MONITORS

Conductimetric Principle

The principle of conductimetry has had wide application as an analytical method for measuring sulfur dioxide in ambient air. In its simplest application,air is introduced into water where the pollutants may ionize and thereby decrease the electrical resistance of the water. The resistance of the solution passing over a pair of electrodes is measured. Because the increase in conductivity is approximately proportional to the ion concentration in the sample, a measure of the ambient pollutant concentration is obtained. This general procedure has been applied in continuous and recording instrumentation for more than 50 years.

Monitors employing deionized water as a reagent are susceptible to interferences of carbon dioxide, salt, aerosols, hydrogen halides, and basic gases. The addition of an oxidizing agent and a small amount of background acid to the distilled water makes the instrument somewhat more specific. This variation of the reagent system, generally employed in the continuous instrument, reduces the solubility of carbon dioxide and thereby minimizes this interference.

Because of their improved specificity, instruments using either colorimetric or coulometric techniques have replaced the conductimetric technique in the past few years in the United States.

Colorimetric Principle

In colorimetry the quantity of a colored component is determined by measuring the relative amount of absorption of light passing through a solution of the desired constituent. In general, chemical compounds are reacted with the pollutant in a predetermined manner under fixed conditions to produce a colored compound. The absorption of light by the dye is related to the pollutant concentration.

This principle is commonly employed for the continuous measurement of three air pollutants - sulfur dioxide, using the West-Gaeke technique; nitrogen dioxide, by the Saltzman method; and oxidants, using the neutral KI procedure; or one of the many variations that exist for each method.

Coulometric Principle

Coulometry measures the quantity of electrical current required to oxidize or reduce a desired constituent. Commercial instruments for the continuous measurement of oxidants, sulfur dioxide, and nitrogen dioxide using coulometry have been developed. This principle had its first application in the continuous monitoring instruments for total oxidants.

To measure the pollutant using coulometry, sample air is passed into an electrolytic detector-cell containing a halide. The free halogen liberated by the oxidants is reduced at the cathode of the cell causing a current flow through an external circuit. The current flow is proportional to the amount of halogen liberated and, in turn, to the oxidant entering the solution.

Measuring the reducing properties of sulfur dioxide involves titration by an electro-generated halogen. A low concentration of the halogen and halide is maintained at equilibrium within the cell by means of a competitive reaction with the cathode and anode. Upon introduction of reactive material, the halogen-halide equilibrium is shifted. The system is returned to its equilibrium condition by a third electrode that regenerates the depleted species. The current required for its generation is measured and is directly proportional to the concentration of the depleted species, which, in turn, is proportional to the quantity of the desired constituent.

This type of coulometric monitor, like the conductimetric monitor described before, is subject to interferences. Materials that undergo oxidation will appear as negative interferences; materials that undergo reduction will appear as positive interferences. Most commercially available coulometric systems are made specific by the use of prefilters that retain interfering compounds and permit passage of the desired constituent.

Recently, three additional detection principles - flame photometry, flame ionization, and chemiluminescence - have found application in automated monitoring systems.

Flame Photometric Principle

Commercially available flame photometric instruments are based on the principle that, when sulfur compounds in air are introduced into a hydrogen-rich flame, they emit light between 300 nanometers and 425 nanometers. The photomultiplier tube views a region above the flame through a narrow-band optical filter to obtain the desired specificity for total sulfur. Sulfur compounds other than sulfur dioxide must be removed prior to entering the instrument in order to obtain valid sulfur dioxide values.

Flame Ionization Principle

Instrumentation employing flame ionization detectors, originally designed for gas chromatographic detection of organic compounds, has had wide application as a continuous monitor for total hydrocarbons. Hydrogen gas, combustion air, and sample air are introduced at a steady, controlled rate before the gases reach the burner tip, where they are ignited and burned in the detector cell. When an organic compound is present in the air stream, the flame produces ionized carbon atoms that result in an ion flow between the flame and the collecting electrode; this small flow of current is amplified and measured. The response for total hydrocarbons is linear.

Chemiluminescence Principle

The chemiluminescent reaction that occurs between an appropriate reagent and the species to be measured produces light energy that is detected by a photomultiplier tube. The resulting current is amplified and, depending on the system, is recorded or, where necessary, electronically stored for further processing. In continuous air monitoring instruments, chemiluminescent reactions of ozone with a dye, Rhodamine-B, or more recently ozone with ethylene are employed to measure ambient ozone concentrations; nitric oxide (NO) with ozone is used to measure nitric oxid concentrations.

AMBIENT AIR MONITORING TECHNIQUES FOR GASEOUS POLLUTANTS

All the previously named principles have found wide applications in continuous air monitoring instrumentation. To avoid neglecting a methodology of general interest, it may be appropriate to summarize available techniques by selected pollutants and to highlight some of the more important characteristics of each method.

Sulfur Dioxide

In the United States a colorimetric principle - the modified West-Gaeke, also known as the pararosaniline procedure - has been adapted as the reference method for measurement of the air quality standard. The reference technique described is of the integrating type and is applicable for sampling periods of up to 24 hours within a range of 25 to 1050 $\mu g/m^3$ (0.01 to 0.40 ppm). The basic technique has also been recommended as the comparison method to the World Health Organization (WHO) by the WHO International Reference Center.

Sulfur dioxide is absorbed in a dilute aqueous solution of potassium tetrachloromercurate (TCM), forming a dichlorosulfitomercurate ion which then reacts with pararosaniline and formaldehyde to form red-purple pararosaniline methyl sulfonic acid. The absorbance of the solution is measured spectrophotometrically.

Major advantages of the reference method are specificity, sensitivity, adaptability over wide ranges, and relative low cost. The methodology lends itself to partial or full automation.

Results of a collaborative test based on the analysis of pure synthetic atmospheres using the 30-minute sampling procedure were: (1) the standard deviation within a single laboratory varied linearly from 15 $\mu g/m^3$ at zero to 36 $\mu g/m^3$ at 1000 $\mu g/m^3$ and (2) the standard deviation between laboratories varied linearly from 29 $\mu g/m^3$ at zero to 70 $\mu g/m^3$ at 1000 $\mu g/m^3$.

The modified West-Gaeke method is readily adaptable for obtaining samples from a large number of stations because the dichlorosulfitomercurate complex is stable before exposure to sulfur dioxide. After exposure, the sample can be stored in a central laboratory at 22°C with a sulfur dioxide equivalent loss of 1 percent per day. When samples are stored at 5°C, no detectable losses of sulfur dioxide occur for 30 days. The analysis can be readily automated to enable a technician to analyze 200 samples per day.

The basic technique has been adapted for automated systems; however, the instruments have relative long response times. Response time is defined as the time interval between a change in the input concentration at the inlet of the instrument to a recording of 95 percent of the ultimate recorded output.

Automated conductimetric methods for sulfur dioxide are simple and inexpensive to operate and have various response times from a few seconds to several minutes. Use of the method in the United States has diminished to a point of insignificance, however, because of its great drawback, the lack of specificity.

Continuous flame photometric instruments are simple to operate, can be supplied with either logarithmic or linear output, have extremely short response times, and are specific for sulfur compounds; with the appropriate filter they are claimed to be specific for sulfur dioxide. The instruments, however, require a source of hydrogen, which presents a potential safety hazard.

Because the reagent is self-generating, continuous coulometric instruments are rather versatile and range in design from very simple to quite sophisticated. One commercially available instrument permits zero and span checks either on a present schedule or by remote command. Because this instrument was designed, developed, and tested here in our host country, det details are not necessary. It should be pointed out, however, that this instrument is the only sulfur dioxide device presently commercially available that has as a standard feature, its own sulfur dioxide source, to enable verification of its operational status. The response times of coulometric instruments are intermediate, ranging up to 10 minutes. Their greatest drawbacks are inherent interferences, i.e., response to oxidizing and reducing species, that have to be removed.

Monitoring systems employing electrochemical techniques are available, but are not commonly used. The sensor consists basically of an electrochemical cell that is covered with a membrane having selective permeability for sulfur dioxide. The transducer converts the pollutant concentration to a current signal by electro-reduction. The signal output of the cells is essentially linear and is insensitive to small flowrate changes. The greatest drawbacks of available instruments are low sensitivity, poisoning, and limited shelf life of the cell. Their advantages are low cost and portability.

Gas chromatographic techniques have been employed to permit separation of parts per billion levels of sulfur dioxide, hydrogen sulfide, methylmercaptans, and carbon disulfide. This mode of separation followed by flame photometric detection of the separated sulfur compounds has led to the development of new types of semicontinuous monitors for sulfur-containing gases that are specific and free from interferences.

Hydrocarbons

Methods for analysis of total hydrocarbons are not numerous. The original flame ionization technique developed for gas chromatography was adapted for total hydrocarbon analysis. The technique employing flame ionization detectors (FID) is essentially a carbon atom counter with a response that is approximately proportional to the number of organically bound carbon atoms in the sample. As a result, FID data are usually expressed as the calibration gas used, i.e., ppm of carbon as methane. The value of data obtained is severely limited because of the high background levels of naturally occurring methane;therefore, the total hydrocarbon method was modified to enable determination of the ambient non-methane concentration. The technique utilizes two instruments: one measures the total hydrocarbon concentration; the other, using a charcoal scrubber preconditioned with methane, becomes a "methane-only" analyzer since the column absorbs non-methane hydrocarbons. The two analyzers are operated in parallel measuring total hydrocarbon and methane concentrations. By obtaining the difference the non-methane hydrocarbon concentration can be calculated.

Recently, instrumentation capable of continuously measuring ambient non-methane hydrocarbon levels by using gas chromatography (GC) coupled with flame ionization detection has become available. The instrument contains a stripper column which removes interfering substances and passes carbon monoxide and methane to an analytical column for separation. The methane is measured directly; the carbon monoxide fraction undergoes conversion to methane in the presence of hydrogen in a controlled-temperature reactor containing a nickel catalyst and is subsequently measured by the flame ionization detector. Direct total hydrocarbon measurement is made without chromatographic separation. Total hydrocarbons less methane is regarded as a measure of reactive hydrocarbons. The gas chromatograph just described is a complex instrument and has only in recent months been improved to the extent that reliable data can be obtained. This technique has been adopted by the U.S. Environmental Protection Agency (EPA) as the reference method for non-methane hydrocarbon monitoring.

Instruments are available with various range combinations. For atmospheric analysis, the common total hydrocarbon range is 0 to 13.1 mg/m^3 (0 to 20 ppm) carbon as methane; the methane range is 0 to 6.55 mg/m^3 (0 to 10 ppm). For special applications, however, lower and higher ranges are available.

A new system for measuring ambient non-methane hydrocarbon levels employs a dual flame ionization detector. One detector receives directly one portion of a split ambient sample for total hydrocarbon determination. The other detector receives a portion of the sample after it has passed through a converter that selectively combusts all hydrocarbons other than methane to water and carbon dioxide. The instrument provides a continuous analog output for methane, non-methane, and total hydrocarbons.

Another instrument also obtains the non-methane concentration by a subtraction process; however, the instrument is cyclic in operation and utilizes only one flame ionization detector. The system analyzes a sample volume for total hydrocarbons followed by analysis of the methane constituent separated by a chromatographic column. The signal for the non-methane concentration is updated every 3 minutes. Another available instrument features automatic standardization of the detector output during each cycle to insure a stable baseline for long periods of operation.

Carbon Monoxide

The infrared absorption characteristics of several gases and vapors make possible their detection and analysis in a continuous analyzer. Non-dispersive infrared (NDIR) analyzers are the most commonly used automated devices for determining ambient carbon monoxide. The method utilized by these continuous instrument systems is based on the absorption of infrared radiation by carbon monoxide. Most commercially available instruments include a hot filament source of infrared radiation, a rotating sector or chopper, a sample cell, a reference cell, and a detector. Generally the detector senses pressure changes on either side of the diaphragm separating portions of gases being irradiated through the sample cell or the reference cell. These pressure changes are converted to electrical signals and amplified to provide an output signal.

Carbon dioxide and water vapor interfere in the determination of carbon monoxide by the non-dispersive infrared technique. Filter cells and optical filters are utilized to minimize these interferences; the use of an air dryer to cope with high absolute humidities is often necessary. Available instruments have fairly fast response characteristics, on the order of less than minutes, and can be purchased to include automatic zero and span capabilities.

The NDIR methodology with a measuring range of 0 to 58 mg/m^3 (0 to 50 ppm) has been accepted as EPA's reference method. The technique has also been recommended by the WHO International Reference Center as the comparison method for WHO.

The NDIR method was tested by means of a collaborative test involving 16 laboratories. The standard deviation within laboratories at the 95 percent level was 1.6 mg/m^3. The standard deviation at the 95 percent confidence limit between laboratories varied nonlinearly from a high of 4.3 mg/m^3 in the concentration range of 0 to 60 mg/m^3 to a minimum of 2.3 mg/m^3 at a concentration of 20 mg/m^3. The tests revealed drying agents and refrigeration methods that satisfactorily reduce the water vapor interference. The use of narrow-band optical filters alone may not provide adequate compensation.

Recently, two instruments – one utilizing dual-isotope-fluorescence, the other an electrochemical cell – have been introduced on the market. At this time no evaluation reports are available.

Nitrogen Dioxide

Air surveillance for nitrogen dioxide concentrations poses a special problem because results of recent studies have necessitated a complete re-assessment of the methodology employed in the United States. The technique employed for many years has been the so-called Jacobs-Hochheiser method, which involves the collection of nitrogen dioxide in an aqueous 0,1 N sodium hydroxide solution to form sodium nitrite. The nitrite ion is reacted with sulfanilamide and N-1-naphthylethylenediamine to form an azo dye that is measured colorimetrically.

Detailed investigations show that the collection efficiency varies nonlinearly from 15 to 70 percent and reveal a positive interference caused by nitric oxide (NO). Consequently, a search for a suitable replacement is actively being conducted. The three techniques primarily considered are: (1) continuous colorimetric method, (2) an improved intermittent (integrating) colorimetric method, and (3) a continuous chemiluminescent method.

The continuous colorimetric method is based on the Griess-Saltzman method. Ambient nitrogen dioxide is converted to nitrite ion (NO_2^-) that, in turn, reacts with a diazatozing-coupling reagent to form a deeply colored azo dye whose absorbence is measured continuously. The method is applicable to measurement at concentrations from 18.8 μg/m^3 to 1880 μg/m^3 (0,01 to 1 ppm). The greatest disadvantage of the method is the inherent negative interference of ozone when the ratio of ozone to nitrogen dioxide is 2:1 or greater.

The integrating method under consideration is the so-called arsenite method with orifice bubbler (ASOB). The method is essentially a modification of the original Jacobs-Hochheiser method. The sample is collected in a dilute aqueous solution of sodium arsenite and sodium hydroxide; chemical analysis follows the conventional Jacobs-Hochheiser method. The method has a constant collection efficiency of 85 percent over the range of ambient levels of nitrogen dioxide. The range of the method is 5 to 740 μg/m^3 of nitrogen dioxide. The precision is approximately 6 percent at 100 μg/m^3.

The chemiluminescent method under consideration is based on the reaction of ozone with nitric oxide at reduced or near-atmospheric pressure. Nitrogen dioxide (NO_2) is measured after conversion as nitric oxide (NO). In the system the ambient nitric oxide concentration is first established, then a switching valve directs the ambient air through a converter where conversion of nitrogen dioxide to nitric oxide takes place. By subtracting the nitric oxide signal from the nitrogen oxides (NO_x) signal, the amount of nitrogen dioxide is determined. The method is applicable to the measurement of nitrogen dioxide at concentrations from 9.4 to 18,800 $\mu g/m^3$ (0.005 to 10 ppm). The precision of this technique is greater than 9.4 $\mu g/m^3$ (0.005 ppm) at the most sensitive range. No interference from pollutants commonly found in the ambient air is known.

Oxidants Corrected for NO_2 and SO_2 (Ozone)

The chemiluminescent methodology is based on the reaction of ozone with ethylene to form a short-lived ozonide that decomposes to form, principally, two molecules of formaldehyde. The chemiluminescent energy released is measured as previously discussed.

Several satisfactory chemiluminescent instruments are available. They display multiple response times, demonstrate great sensitivity, provide excellent zero stability, and have no known interference. Their disadvantage is the requirement of a supply of ethylene. Prior to use, the system is standardized against the neutral phosphate-buffered potassium iodide method. The chemiluminescence principle has been selected by EPA as the reference method to determine air quality for oxidants corrected for nitrogen dioxide and sulfur dioxide. The range is from 9.8 $\mu g\ O_3/m^3$ to greater than 1950 $\mu g\ O_3/m^3$ (0.005 ppm O_3 to greater than 1 ppm O_3).

At least one commercially available instrument measures ozone continuously by detecting the absorption of ultra-violet light at 2537 Å. Specificity for the measurement is obtained by alternately sampling ambient air and air from which ozone has been removed; the difference is a measure of the ozone concentration in the gas sample. Therefore, the ambient air is sampled and measured in continuous cycles.

Continuous colorimetric analyzers usually employ a 10 to 20 percent solution of neutral-buffered potassium iodide solution as the absorbing medium. The iodide is oxidized to iodine and tri-iodide ion, which are determined colorimetrically at 3520 A. Sulfur dioxide, a serious negative interference, can be reduced by utilizing a prescrubber of chromium trioxide (CrO_3). Measurement is further complicated by the fact that NO_2 is a positive interence. The CrO_3 scrubber adds to the problem by oxidizing nitric oxide to nitrogen dioxide necessitating an additional correction factor. Compounding these problems is the fact that CrO_3 scrubbers are moisture sensitive and display varying efficiencies.

Coulometric methods have also been used extensively for the measurement of oxidants. In this technique, the ambient air is introduced into an electrolytic detector-cell containing potassium iodide. As mentioned in the discussion of coulometric methodology, the current flow caused by the reduction of free iodine at the cathode is proportional to the oxidants entering the solution. Depending on the size of the cell, the response time can be quite short. Instruments and operating costs are low; however, the technique lacks the desired specificity.

AMBIENT AIR MONITORS FOR PARTICULATE MATTER

Particles Larger Than 10 Microns

Particles larger than 10 microns have in the past been collected in dustfall jars. Considering all factors including expenses, it appears that no measurement technique presently used to determine the concentration of particles larger than about 10 microns is superior to a properly installed dustfall jar. Unfortunately, the measurement generally does not correlate with emissions.

Another device that has found application is a cylindrical container with adhesive coating on the outside. It was used to ascertain the wind direction corresponding to the maximum dust content of the air, which often permits identification of major dust sources. In Europe, the use of flat adhesive surfaces placed horizontally as dust collectors was the favored practice.

Particles Less Than 10 Microns

Particles in the size range from 0.1 to 10.0 microns in diameter constitute the greatest fraction of the atmospheric particulate mass. The greatest optical effects also arise from particulates in this portion of the size spectrum. One simple optical technique involves the use of a photometer for determing air turbidity. Another more complex method employs the nephelometer, which measures light scattered by particles suspended in a defined volume of air. Other light scattering instruments include total forward scattering and right-angle scattering photometers as well as instruments that count and size individual particles.

In the United States, cascade impactors have been found to be useful for simultaneously collecting and classifying particles throughout most of the 0.1 to 10.0 m size range. Cascade impactors yielded considerable information on the urban aerosol size distribution of selected chemical components including sulfate, lead, and other metals.

In spite of the availability of these methods for measuring particulates in air, however, it has been somewhat routine practice in the United States to measure suspended particulates in the atmosphere by measuring the change in optical density as a result of filtered deposit or by using a gravimetric process.

The commonly used method in the United States employs the AISI (American Iron and Steel Institute) sampler. Other versions exist, but all are alike in function. A series of portions of filter paper, usually successive areas of a paper tape, are clamped between an intake tube and a vacuum connection. Air is drawn through a filter for a selected time, a new portion of tape is then moved into position, and sampling is resumed.

The fundamental basis of evaluating the collected samples is optical. Reflectance may be measured photometrically, or the transmittance of light through both filter and deposit may be compared with the transmittance through a clean portion of the filter. Reflectance measurement determines the blackness of the deposit; transmittance measures a function of all particles collected. Generally speaking, reflectance and transmittance correlate fairly well over short time periods. Over longer periods the introduction of new sources and the removal of old ones may be expected to change the composition of the particles enough to cause divergent results between the two techniques.

The optical density relationship measured by transmission, which is the most widely used technique in the United States, is normally converted into units of coh's per thousand linear feet of air passing through the filter. The unit, coh, (coefficient of haze) is defined as the quantity of particulate matter that produces an optical density of 0.01 when measured by light transmission. A coh measurement is routinely reduced to coh per thousand linear feet of air passing through the filter tape by dividing the inreduced coh value by the number of thousands of feet actually drawn in the test.

For reflectance, a "ruds" test is made by measuring the percentage reflectance of the filtered tape and reducing the measurement of 1000 linear of air. Reflectance of clean paper tape is the reference standard, set at 100 on the reflectometer. Ruds is an abbreviation for "reflectance unit of dirt shade" and is used in various European countries for calculating concentrations of "smoke" or "dark suspended matter".

The EPA reference method for the determination of suspended particulates in the atmosphere is the high-volume method. The design of the high-volume sampler has not changed substantially since its conception and does not need to be reviewed. Conventional high-volume samplers employ a horizontal filter surface protected by a shield or roof, which keeps off rain and snow and generally prevents collection of particles larger than about 100 microns in diamter. Filters are made of glass fiber with a collection efficiency of at least 99 percent for particles of 0.3 microns diameter or greater.

The recommended flow rates of 1.132 to 1.698 m^3/min (40 to 60 ft^3/min) allow the suspended particles having diameters of less than 100 microns (Stokes equivalent diameter) to pass to the filter surface. Particles within the size range of 0.1 to 100.0 microns in diameter are ordinarily collected. The filters are equilibrated prior to weighing for 24 hours at a temperature of 15 to 35^{0}C and a relative humidity of less than 50 percent both before and after sampling. Samples are normally collected for 24 hours and sampling rates are measured at the beginning and end of the period.

The collected sample represents the particulate content of approximately 2000 m^3 of air and is large enough for almost any type of analysis. If detailed analysis of the particulate is contemplated, care must be exercised to use filters that contain low background concentrations of the pollutant being investigated. Careful quality control is required to determine background values of the filter.

In European practice the Generalized International Standard Calibration Curve is commonly used to relate transmittance of the tape spot to the concentration, in $\mu g/m^3$, of the suspended material. The use of the conversion curves to make international comparisons may be potentially dangerous, however, because the method does not state that actual airborne particulate concentrations have not been determined gravimetrically. This is of concern because air quality standards in the United States are stated in gravimetric units, and the reference method used to determine compliance with the standard is the high-volume sampler.

Notwithstanding these limitations, the relative values of tape samplers should not be overlooked. Tape sampler results should, however, be used with the knowledge that the measurement in whatever units is arbitrary, artificial, and without absolute meaning. The gravimetric method is, therefore, the recommended method for determining ambient air quality standards for particulate matter.

A collaborative test of the high-volume method by 12 participating laboratories provided the following results: (1) the relative standard deviation within a laboratory was 3 percent; (2) the relative standard deviation between laboratories was 3.7 percent; and (3) the minimum detectable amount of particulate matter was 3 mg, which is equivalent to 1 to 2 μg/m³ for a 24-hour sample.

The WHO Meeting on Analytical Methods and Data Processing in Air Pollution Monitoring recommended that the international comparison method should be gravimetric and be based on an improved version of EPA's high-volume sampler.

New Instrumentation

A relative new device, the piezo-electric sensor, delivers real time measurements over a particle size range from 0.01 to 15 m at concentrations up to 1000 μg/m³. The sensor is applicable to monitoring the ambient atmosphere,stack effluents, and motor vehicle emissions. Air to be tested is drawn through the unit by a small accessory vacuum pump. As the gas enters a Teflon chamber, particles are electrostatically charged. The flow is directed over a piezo-electric sensor and an electric field forces all of the particles into contact with the sensor. The adhering particles lower the residence frequency of the sensor in direct proportion to their mass. A second piezo-electric sensor mounted in the chamber serves as a reference for the sensor. The measuring and reference sensors are connected in a bridge circuit, so that any change in the relationship between the frequencies of the sensors will be caused only by particulates accumulated on the measuring sensor. The output signal of the device is generated by measuring the difference between the measuring and the reference sensor frequency. The rate of change of this output frequency is directly proportional to the mass concentration of the particles.

Another device, the beta absorption tape monitor, has also recently been used in the measurement of total suspended particulate. In this monitor, particulate matter is deposited on paper felt tape for a preset period of time. The tape spot is continuously exposed to a stream of beta particles from a small radioactive source. The beta particles interact with the sample and the decrease in the beta particle count sensed by the detector is directly proportional to the number of electrons present in the sample. Thus, the technique measures a parameter that is directly proportional to mass.

CONSIDERATION FOR AUTOMATED MONITORING SYSTEMS

A problem commonly encountered with ambient air monitoring is the absence of data on instrument performance in the field. One of the reasons for this dilemma is the proliferation of available instruments. More than one hundred different models of continuous instruments are available in the United States. A related problem faced by air pollution control agencies is the selection of a complete system.

Instrumentation Evaluation

Despite many attempts to evaluate instrumentation, the sheer number of systems and their constant modification make it almost impossible to obtain an objective judgment of the merits of a given instrument, two sets of data must be available. The first set should include data obtained through evaluation of the instrument in the laboratory, such as performance specifications determined under carefully controlled test conditions. The second set of data should be obtained under field conditions over long periods of time and should include maintenance, calibration, and repair requirements.

All too often in the past, instruments with favorable evaluation data obtained under laboratory conditions proved to be extremely difficult to operate in the field. The reasons for this problem are numerous, complex, and beyond the scope of these introductory remarks. It seems appropriate, however, to mention the primary causes of the problem, which arise because of the different conditions under which the instruments are operated.

In a laboratory, the evaluation is carried out by highly skilled scientists under well controlled conditions and is performed on a single instrument for a limited time period with a high degree of interest and support displayed by the seller.

For most agencies who are collecting ambient air quality data, the conditions under which they must operate are quite different. The operators are technicians having minimal training or no training on the specific instrument at all and thereby having to rely on a possibly inadequate instruction manual. The technician also works in a shelter without environmental controls and without the aid of the vendor.

One of the greatest misconceptions existing in this area is the notion that an air monitoring instrument can be installed and calibrated in the field without too much difficulty and will from then on operate perfectly and generate valid data with only occasional attendance. Nothing could be further from the truth. Problems encountered during operation of atmospheric analyzers include the establishment of stable flow rates, effective gas-liquid transfer, stable electronic responses, and proper quality of reagens and/or gases. To assure the validity of data, a vigorous quality control and maintenance program is required. Proper calibration is a prerequisite for any measurement. The interval between instrument calibrations of continuous air monitors is also of major importance because some malfunctions are not evident from daily observations and are only discovered during calibration. Air quality data is often subject to criticism because periodic calibration data is not available. Successful ambient air surveillance requires an effective quality control program and sufficient properly trained personnel to carry out the program.

Availability of accurate monitoring techniques is a necessary first step for the control of air pollutants and the setting of guidelines or standards.

A complicating factor in dealing with various competing instruments is that they may vary widely in their specificity and other performance characteristics. In 1968 the National Air Pollution Control Administration conducted both laboratory and field instrument comparison tests. The major results of the tests were twofold. First, users became aware of the variation in performance specifications in the instruments. Second, manufacturers also became aware of the discrepancies and made some necessary modifications.

Presently, EPA is actively considering a mechanism by which a given method and/or instrument can be declared equivalent to the promulgated reference methods. It is envisioned that the burden of proving that a candidate method is equivalent to a reference method rests with the requesting party. The requesting party must demonstrate that the candidate method meets certain performance specifications and/or generates valid air quality data that have a consistent relationship with the reference method.

Selection of Automated Monitoring Systems

To fulfill the monitoring requirements for a given task, the objective must be clearly stated in sufficient detail to permit selection of a system based on the specific requirements. A frequently posed question is the one asking what type instrument to procure. A proper analogy would be to ask an automotive engineer what vehicle one should buy. Obviously, without the proper background information the question cannot be answered. It seems proper, therefore, to examine briefly two of the more obvious parameters that are often overlooked.

The first is the selection of the sensor. To make an intelligent decision, answers to the following have to be provided:

1. What instrument range will cover the concentrations under investigation?
2. What are the required averaging times such as minutes, hours, or days?

3. What type of personnel is available to operate and maintain and/or calibrate the system?
4. What other pollutants are or may be present that could interfere with the methodology employed?
5. Are laboratory or repair facilities available?

Without answers to these extremely important questions selection of instrumentation is virtually impossible.

The second parameter deals with choosing a recording system. Traditionally, continuous air monitoring instruments have been used with a good quality strip chart recorder. The strip chart recorder has certain advantages that cannot be replaced by the most sophisticated data acquisition system.

The advantages of a strip chart recorder are briefly summarized as follows: (1) it provides a visual record of past performance of the instrument; (2) daily zero drift corrections, which are generally more serious than the span drift, can be made and hourly average estimates can be obtained every day; (3) a permanent record is provided permitting supervision of the operator and a daily validation procedure using original data; and (4) provides evidence in courts if needed. Of course, there are disadvantages, the greatest being the manpower requirement. In the United States, however, there is an economic advantage in utilizing volunteers and young people for this less sophisticated task rather than employing highly trained personnel to keep a fully automated system operating properly.

Semi-automated, intermediate systems such as magnetic tapes or punch tapes can be utilized avoiding the tedious task of reading and averaging strip charts. In my judgment, however, a strip chart is still required to enable final validation of the computer print-out against an original record.

One has to remember that regardless of the recording device being utilized the validity of the data depends on the sensor. The sensors, at least in the past, have not been sufficiently reliable to accept the generated data without questioning. At the present time, telemetering and data acquisition systems should be used with extreme caution and as the exception and not the rule. However, in the age where live televised pictures from the moon can be received into the home, the desire to have a so-called real time system telemeter data to a central facility is understandable. It is obvious that a more logical approach is required.

A telemetering system with a central data acquisition system should not be considered unless dictated by scientific or economic factors. Most present day instruments have drift characteristics that require almost daily corrections, although instrument performance has improved recently. We are rapidly approaching the time when reliable data can be obtained over a 2 to 3 day period. If a requirement to visit an instrument or an automated system approximately once a day to accomplish daily zero and span functions has to be included in the instrumentation package, then telemetering becomes less attractive unless a large number of stations are being operated in the same general area.

Last but not least one has to face the difficulties of selecting a site and shelter. Selection of instruments and recording devices should, if at all possible, not be dictated by site selection criteria. Unfortunately, site selection is often dictated by the availability of suitable shelters and the suitability of the shelter depends on the chosen instrument. Site locations are extremely important, especially if it is anticipated that the station will be in operation for a long period of time.

Relevance of the obtained measurement depends on choice of the sampling site; therefore, it is of utmost importance to keep the objective of the study in mind. Not only meteorological parameters but also present and future single source influences must be considered. If one should be fortunate enough to find an ideal location, then one has to be concerned with the proper shelter.

The shelter may be a prefabricated unit, a mobile laboratory, a public service facility, or space rented from a private source. Our experience has been that public facilities such as a fire station, police station, school, or hospital make excellent sites for the installation of continuous monitoring system if not influenced by a local pollution source. There are at least three important considerations in selecting the space required: (1) access to the site is necessary for every day of the week, (2) adequate electrical power has to be available (Many times this will necessitate an installation of extra circuit.), and (3) the room has to have reasonably constant temperature.

In addition, secondary requirements such as protection from vandalism and safety of inhabited buildings are required because many newer systems require the use of flammable or explosive gases.All of these considerations have an influence on the selected automated monitoring system.

Consideration has to be given to the location of the sampling probe intake with respect to the presence of any obstructions that may influence a concentration of the ambient air concentration prior to entering the intake. The problem of where the sample intake should be located with respect to sources and structures has not been satisfactorily standardized. A classic example of a pollutant very sensitive to site and intake probe location is carbon monoxide.

Furthermore, consideration must be given to the design and construction of the manifold to assure delivery of a representative sample to the analyzer. To prevent interaction of the pollutant with the manifold, an inert material such as glass or teflon should be used. One preference for glass is because the accumulation of foreign substances, for example water and particulates, can be observed and remedial actions can be taken.

ENVIRONMENTAL POLLUTION CONTROL SYSTEM AND THE PERIPHERAL DEVICES

Y.KUMAZAWA

Environmental Pollution Control Center, Osaka Prefecture, Japan

INTRODUCTION

The environmental pollution control system, consisting of telemetering and computer systems, conducts continuous monitoring and takes emergency measures. The telemetering system transmits the measured data from individual monitoring stations to the central surveillance room in the Environmental Pollution Control Center by wire or wireless, and the computer system functions are to arrange,collect, summarize, maintain, and store these data and to display or print them out.

The system was first introduced to effectively execute the continuous monitoring emergency actions especially for air pollution, for which the local government is authorized by the Air Pollution Prevention Law. At present the system is being enlarged to include the continuous watch for sources of air pollution, water pollution, and noise pollution. The scale and activities of the system, however, differ according to the local pollution condition in the prefecture and to the budged appropriated.

CONTINUOUS MONITORING

The purpose of continuous monitoring is to keep observing and watching the condition of air pollution, and naturally involves making comparison of the measured data with the national standards, the investigation of yearly change and the trend of pollution and of pollutants from specific sources and the inter-local diffusion of pollutants. An accurate and immediate transmission is required from the pollution telemetering system, allowing no later measured data from the chart-paper, since this work must be done in real time with on-line processing. Accordingly, the continuity and reliability of the auto-analyzer at each station is most necessary for constant monitoring. All analytical determinations must be made at the monitoring stations stationed at local places by the method designated by the Board of Environment of the National Government. Meteorological conditions such as wind speed, wind direction, temperature, humidity and solar radiation are also determined simultaneously at the monitoring stations. The determined data are transmitted to the central telemetering receiver connected with the computer system every 10 minutes and displayed on the sign board or printed out by the high speed printer every hour.

EMERGENCY ACTIONS

When the concentrations of pollutants exceed the emergency level, the emergency order is issued to ask or order the factories to change to low-S fuel, or for the curtailment of operation of the factories by means of alarm receivers installed in the 200 designated factories, as well as to the city offices and inhabitants in the prefecture to avoid the toxic pollutants. For this emergency order, the central instruction room must be kept fully informed on the conditions of air pollution throughout the emergency time. Consequently, the continuous monitoring and the emergency actions are closely related to each other, and this is the reason why the immediate transmission by telemetering is necessary for this system. For the smooth performance of the latter, the reliable operation of the former is vital. Moreover, the measured data must be transmitted continuously and herein lies the importance of the telemetry transmission. The pollutants requiring the special activities on emergency designated by the Environmental Pollution Protection Bureau are the five following: sulfur dioxide, airborne dust, nitrogen dioxide,oxidant and carbon monoxide. While the emergency order is being issued, the responsible officials of the prefectural government go around the factories by patrol cars equipped with telemetering devices to check how the factories are carrying out their pollution abatement activities.

UTILIZATION OF INFORMATION

Continuous monitoring is not only vital for the inspection of environmental conditions, but also makes accumulated data available for various statistical purposes. Data produced by the constant monitoring are indispensable for the statistical investigation of damage caused by pollution, the evaluation of the effectiveness of control measure and comparison with other local prefectures on the pollution situation. These data, therefore, must be stored in memories so that they may be easily picked out whenever necessary or be exchanged automatically among prefectures, as well as between the prefecture and the national government. On account of this, various plans are being carried out for constructing the data bank.

ACTUAL ACTIVITIES REQUIRED FOR THE ENVIRONMENTAL POLLUTION INSPECTION

Continuous monitoring is work for the long-term betterment of the environment, and emergency actions for the short-time abatement to cut out the high concentration of pollutants. These works are accompanied by the following actual activities.

Information for the public

The basic principle is for the data on environmental pollution to be made known to the public, and sometimes in real time. Consequently, the information is displayed on electric sign boards installed at busy streets. At present, weight is placed on public information on environmental pollution rather than the display of the data. But this center is often approached by people asking for data by telephone and letters. Informed by publicity they often want to know the facts in more detail. Good maintenance of the equipment is necessary so that it can indicate correct data. The measured data are printed out every hour to be used for answering inquiries from the public as well as for judgment at the time of an emergency order.

Forecasting

On emergency, every factory is requested to change to non-polluting fuel or to reduce combustion or to curtail operations. However, these emergency measures place a considerable economic burden on the enterprises. So the fewest emergency orders are more desirable for them. It is necessary for the control administrator to predict a high concentration of pollution in order to give the enterprises the minimum economic burden, and to ask the factories to be ready for the emergency abatement in advance. The forecast is made by judging from the telemetered data, past and present and the weather forecast sent from the central meteorological observatory to the center by wireless. On the other hand, a prediction method for high pollution levels is being studied and developed for practical use. The moving mean values for the past 48 hours, the change in pollution level at every 10 minutes and other data necessary for forecasting are easily printed out by the one-touch operation to be used for forecasting work.

Mass Communication

Recently the mass media have been very much interested in information on environmental pollution. Information such as the records of emergencies, the maximum concentration of pollutants in the past, the number of people suffering from photochemical smog as well as pollutant concentration in emergencies, must be ready to be picked out whenever they are asked for. When requested by a reporter, this information must be immediately printed out from the data bank system in a uniform style by means of specifically provided program. In local government assemblies the counter-measures for environmental pollution are earnestly discussed. So at any inquiry by assembly members, quick provision of exact information is necessary. However, the large scale computer system is necessary to answer the specific inquiries raised in assembly discussion, for they want large scale measurement.

Monitoring Emission Sources

It has become the dominating idea that the fundamental concept of preventing environmental pollution is to control the emission source before the pollutants are diffused. The data of emission sources at individual factories are also telemetered to the central instruction room for collective control. The subsystem to monitor emission sources will be available for the confirmation of abatement activities on emergency. Besides the estimation of environmental pollution, the future planning as well as the daily inspection of sources will be effectively conducted if the system functions well.

Large Scale Computation

The accumulated data are arranged in daily and monthly reports and conformity to the environmental standard is judged statistically every year. Besides, the large-scale computation such as the estimation of environmental capacity, the long- and short-term prediction of future pollutant concentrations, the detection of pollutant sources, the calculation of diffusion equations and the statistical processing of damage caused by photochemical smog are also necessary. These computations require a large computer organically equipped with a data bank system and an operator.

FUNCTION OF ENVIRONMENTAL POLLUTION WATCH SYSTEM

To conduct the abovementioned functions the following are essential for the computer system.

1. All data transmitted from each monitoring station are processed real-time and on-line, and printed out on the high-speed printer every hour, and shown on the sign board. If necessary, comparison of the data with emergency levels is also indicated.
2. The maximum value, the minimum value, and the mean value are printed out in the standard format at the zero hour of the day.
3. Errors of the analytical instrument and mistakes in the telemetering equipment are checked.
4. The moving mean values for the past 8, 12 and 24 hours are put out arbitrarily.
5. Back up system is provided for saving the data during a breakdown of computer processing.
6. Disorder of the system must be self-diagnosed and the disorder indicated.
7. The data from environmental and source monitoring must be automatically stored in the largest possible memories.
8. Information necessary for the public administration, such as the number of emergency orders by days and total hours, the list of designated factories at a pollution emergency, and emission standards must be stored, if possible, in the data bank.
9. The stored data must be easily added, corrected, and deleted, and quickly picked out and displayed.
10. Inquiries for information made by the press and the prefectural assembly as well as by public information activities must be readily and rapidly answered whenever asked.
11. Information exchange with other prefectural governments and municipalities shall be possible.
12. Various large-scale calculations,future planning, simulation, diffusion, and environmental capacity shall be possible by using the stored data.
13. An alarm receiver shall be installed at each factory designated as one at the time of pollution emergency by the Osaka Prefectural Government and the municipalities, town and village offices so that a message by voice can be received.

When the above-stated functions have been satisfactorily fulfilled, the control center will have a large-scale watch system equipped with a central large scale computer as the core of the system, and CRT (display apparatus), sign board, type-writer, disk, high speed printer, magnetic tape, X, Y ploter, card reader, central console, microphone, standard watch, central telemetering device, magnetic drum, data exchange instrument and card punch as peripheral devices.

PERIPHERAL ANALYTICAL INSTRUMENTS

It is not too much to say that the success of the environmental pollution monitoring system mainly depends on the reliability and reproducibility of the analytical instruments which operate at the terminal monitoring stations. Consequently, the analytical instruments are required to have following features.

Reliability

One of the most important aspects of the reliability of an auto-analyzer is how precisely the instruments respond to the variables of ambient air over a long period. The drift of sampling flow, the shift of zeropoint, and the decline of sensitivity will accompany the auto-analyzer, so that they must be checked frequently for the individual instruments at least twice a week. Otherwise, the data will not be reliable. The less such errors are, the better the reliability.

Durability

The average life of an auto-analyzer is 2 - 5 years. Local governments cannot afford to renovate these instruments every two years. The further development of these instruments must extend the life to more that 5 years. However, as these instruments are usually operated for 24 hours a day through the year, the sensitivities often decrease in the latter half of the 5 years. In such cases some parts may be replaced but such a remedy has naturally a certain limit.

Ease of Maintenance

The renewal of a gas-bomb, refining of drying reagent, replenishment of absorbing solution, renewal of active carbon, recording paper, recording ink and so on are laborious tasks. For instance the deterioration of drying reagent by continuous use sometimes causes a lowering of the sensitivity of an auto-analyzer. In such a case, further development of an instrument to eliminate the use of the drying reagent is better than relying on regular maintenace. Easy maintenace can considerably reduce gaps in measured data. and maintanance twice a week is enough. Development of instruments which will have good sensitivity with one check a week is much desired.

Sensitivity

Analytical sensitivity is also important, but the reliability or durability should not be sacrificed for improvement in the sensitivity. Various attempts have been made to develop new analytical instruments which will be more sensitive for SO_2 and NO_2.

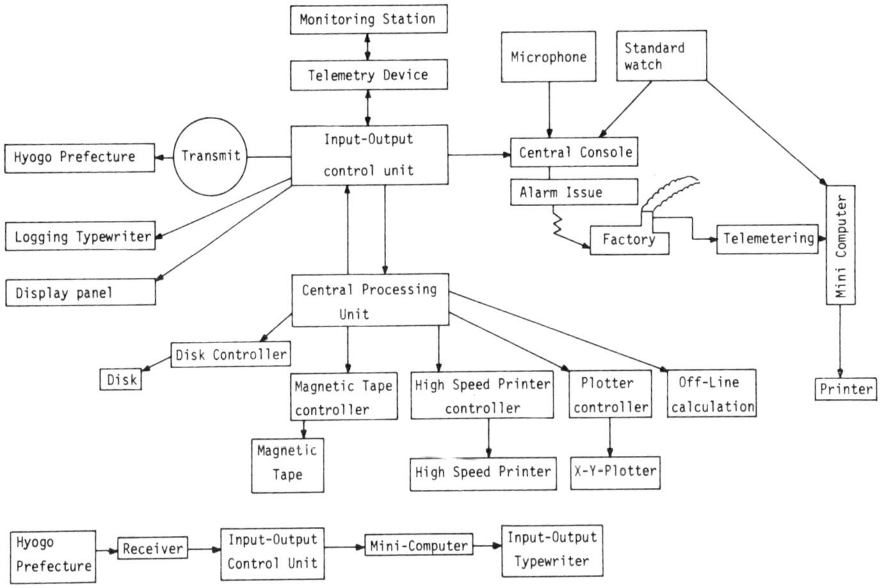

Fig. 1 - Environmental Pollution Watch System

Table I - Emergency level

Pollutants	Forecast	Alarm 1 (Precaution)	Alarm 2 (Warning)	Alarm 3 (Emergency Alarm)
SO_2	Previous Day On the Day	1. 0.2 ppm x 0.5 ppm Continue for 2 hrs 2. x 0.15 ppm Mean Values for 48 hrs 3. Judging from Meteorological Condition after Issuing Fore-cast	x 0.5 ppm	1. x 0.5 ppm Continue for 3 hrs 2. x 0.7 ppm Continue for 2 hrs
DUST		x 2.0 mg/m^3 Continue for 2 Hours		x 3.0 mg/m^3 Continue for 3 Hours
CO		x 30 ppm		x 50 ppm
NO_2		x 0.5 ppm		x 1 ppm
OXIDANTS	x 0.10 ppm	x 0.15 ppm	x 0.3 ppm	x 0.5 ppm

Table II - Control of emission sources on emergency

	Alarm 1 (Precaution)	Alarm 2 (Warning)	Alarm 3 (Emergency Alarm)
Sulfur Dioxide	SO_2 0.2 ppm for 2 Hours	SO_2 0.5 ppm	SO_2 0.5 ppm for 3 Hours SO_2 0.7 ppm for 2 Hours
	40% Reduction of SO_2 Emission (Fuel Change)	50% Reduction of SO_2 Emission (Fuel Change)	70% Reduction of SO_2 Emission (Fuel Change, Operation Control)
Particulates	S.P 2.0 mg/m^3 for 2 Hours		S.P 3.0 mg/m^3 for 3 Hours
	20% Reduction of Total Emission		40% Reduction of Total Emission
Nitrogen Dioxide	NO_2 0.5 ppm		NO_2 1 ppm
	20% Reduction of Combustion Activities		40% Reduction of Combustion Activities
Oxidants	O_3 0.15 ppm	O_3 0.3 ppm	O_3 0.5 ppm
	20% Reduction of Total Emission Gas		40% Reduction of Total Emission Gas

DISCUSSION

JANTUNEN

What action is taken in Osaka when high CO levels occur?

KUMAZAWA

In Japan, we have not yet suitable countermeasures to cut down the traffic flow emitting carbon monoxide even when CO concentrations exceed the emergency level. The authority to control traffic belongs to the police department. However, the department does not easily co-operate with the local government for cutting down the traffic flow when high CO levels occur. The department has the idea that preventing traffic accidents is more important than controlling air pollution. The prefectural government has had frequent meetings with the police department to discuss countermeasures to control air pollution, but in vain. In conclusion, Osaka has no emergency action for high CO concentrations.

DE KONING

What type of work was done to decide that the 10-minute averaging time was good enough for the work you are doing?

KUMAZAWA

According to the National Air Pollution Prevention Law, the emergency concentrations are given as one hour averages. However, the emergency action in Osaka must be taken quickly. Therefore, alarm is often issued or cancelled according to a 10-minute average value. The 10-minute values are used to enable factories to take prompt action. This will take about 30 minutes.

STEWART

How often the last years have the threshold levels been exceeded; how often have emergency procedures been invoked. What is the total response time of the system, including the industrial action response, before a noticeable and worthwhile improvement in the quality of the air results?

KUMAZAWA

On average, we issue precautions 8 to 10 times a year for SO_2 and 10 to 20 times a year for photochemical smog. Referring to your question on the total response time I can only say that it is very difficult to evaluate how much the cooperation of the factories contributed to the diminishing of the pollution concentration.

RUFF

Dr.Burmann mentioned the use of volunteers to analyze strip-chart recordings. How accurately they can do this with reference to the interpretation of thermal and span drift?

BURMANN

We know that they can do it very accurately since we have taken volunteers and have trained them. We have given them unknown strip charts and our statisticians have tested the results to see how well they can dot it. They calculated hourly averages just as well as the instrument can do.

WEBER

If you think of an automatic monitoring station equipped with instruments for SO_2, NO_2, oxidants, hydrocarbons, etc. and it is equipped with the best instruments available, how many people do you need to run the station and what is the educational level of these people?

BURMANN

If you have a well equipped station with the instruments you mentioned, you have no data acquisition system, and you want to have daily zeros and daily spans on each instrument, you will need one person per station. The eductional level need not be too high: a high school graduate can do the job. You have to provide quality control from the central laboratory: the technician does not actually decide if it is good or bad data. If he observed an excess of drift or an excess of a standard he has to call the laboratory.

BERLIN

Dr.Burmann stated that location of sampling probes was most important when assessing the levels of pollution and he indicated that this was most crucial for carbon monoxide. As the frequency distribution of carbon monoxide is very variable, where are the sampling stations really located?

BURMANN

Two years ago we were forced to employ 40 CO-instruments. And at that time the discussions were quite intense: "where are you going to locate these instruments". If your are looking for health effects you are concerned with people. Then you have to site the instruments in places where people live. I would like to see them situated where there are people who are already punished by economics to live in areas which are underprivileged. You are not going to take you sampling probe 30 cm from the curb because all you are doing is monitoring the exhaust of an automobile that is sitting at the intersection. What concerns me greatly is the location of the intake because we have had the experience that just by moving the intake another 2 or 3 m up in the air you will get a completely different set of data. Whatever you do, you have to standardize in the sense: here is the instrument, these are the conditions, and leave the instruments alone. Otherwise you can not get comparable data.

BERLIN

Have you considered the problem of indoor - outdoor ratio measurements? The air quality standards do no specify where the measurement is supposed to take place. How do you really cope with the problem since you mentioned that one of the purposes of monitoring is compliance with standards.

BURMANN

Since the time we were forced to employ the 40 instruments, we have had some directions out of Washington and these are: you will place the instruments at the area of maximum concentration, which is much clearer than a few years ago.

SHEARER

Our agency has recently completed a study in a large apartment building in New York city with a freeway going underneath the building itself. A report has been issued on this. The results were that we found at apartment levels up at the 32nd floor CO concentrations inside and outside the apartment were higher than at street level. And in one apartment on the 32nd floor our 8 h standard was exceeded during 12% of the time.

CURRENT WHO ACTIVITIES IN THE MEASUREMENT AND ANALYSIS OF URBAN AIR POLLUTANTS

GRAHAM CLEARY

Environmental Pollution Division of Environmental Health, World Health Organization, Geneva, Switzerland

INTRODUCTION

The World Health Assembly resolution WHA 24.47 endorsing the Director-General's proposal for a long-term programme in environmetal health stressed *inter alia* the need

> "... to stimulate the development and coordination of epidemiological health surveillance by methods including environmental monitoring systems, in collaboration with other national and international efforts, in order to provide basic information on actual and suspected adverse effects on human health attributable to the environment; and to extend the knowledge of effects of environmental factors on human health by collection and dissemination of information, stimulation, support and coordination of research, and assistance in the training of personnel".

The WHO collaborative programme in air monitoring is implemented through the WHO reference centres and collaborating laboratories on air pollution which include, at present, two international reference centres, the IRC for Clinical and Epidemiological Aspects of Air Pollution at the MRC Air Pollution Unit, St.Bartholomew's Hospital Medical College, London and the IRC on Air Pollution Control at the Environmental Protection Agency, United States of America. In addition, there are three regional reference centres at Moscow, Nagpur and Tokyo, seven national reference centres and eleven collaborating laboratories (plus laboratories in nineteen cities in the Pan American Sampling Network).

One of the main functions of the reference centres and collaborating laboratories is to promote the use of uniform methods of measurement, to introduce reliable and effective procedures for the calibration of routine sampling and analytical methods, and to improve the design of monitoring systems and the handling, statistical analysis and use of data. The programme will also help in generating internationally comparable data on levels and trends of air pollution in some urban and industrial areas; this may help to identify patterns of exposure to air pollution, and facilitate the planning and assessment of health effects studies carried out in different countries, and the comparative evaluation of the effectiveness of national air pollution control programmes.

The principal areas of WHO involvement in air quality monitoring are summarized in Chapter 2, Duration and type of monitoring; Chapter 3, Epidemiological basis for monitoring certain pollutants; Chapter 4, Analytical Methods and Chapter 5, Processing of Air Quality Data.

DURATION AND TYPE OF MONITORING

Because pollutant levels at any sampling site usually fluctuate widely with time, where practical (and economically feasible), the most detailed data for scientific studies is obtained with a monitoring instrument that provides a continuous record. Nevertheless, the long response time of some instruments may pose problems.

So far as non-instrumental methods are concerned, monitoring for control purposes usually involves relatively short periods for sampling. This may be from one or two hours up to a 24-hour period. Shorter intervals are indicated when peaks are likely to occur, when the effect is prompt or when the material may be evanescent.

Monitoring for epidemiological purposes may require similar time spans as monitoring for control, particularly if maximum values are believed to be important. If mean long-term exposures are to be considered, then simpler integrating types of monitoring methods may suffice, the purpose being to develop a base-line or defined exposure level against which changes in the level of pollution can be assessed as well as the effects of such change on the population under study.

The accuracy of results depends upon the method of measurement and the relevance on the choice of sampling site, the time when the samples are taken and the duration of sampling. Sites should be chosen so that the air sampled is representative of the air breathed by the population concerned*. (Bad siting can lead to concentrations that are considerably higher or lower than this and which may in fact have little or no meaning).

---

*This may also involve such aspects as monitoring indoors and personal monitoring in order to assess the total exposure of individuals.

EPIDEMIOLOGICAL BASIS FOR MONITORING CERTAIN POLLUTANTS

Historically, suspended particulate matter and sulfur oxides have been the pollutants most commonly measured. These, either as indices or a causative factor, have been associated with excess mortality when the levels have been high. At lower levels they are associated with increased morbidity and may be important factors in the development of chronic lung disease. At very low levels they may be associated with changes in pulmonary function, the clinical significance of which is not clear at present. It may be that other materials such as sulfuric acid, sulfates and oxides of nitrogen should be measured in conjunction with the measurement of suspended particulate matter and sulfur oxides. All of these different materials might then be examined with respect to their role as possible aetiological factors in the production of disease.

With increasing urbanization and the increased use of petrol-driven motor cars, carbon monoxide has become an important urban pollutant. (Cigarette smoking is an important although often forgotten source.) Inhaled carbon monoxide combines with haemoglobin to form carboxyhaemoglobin (COHb). This interferes with the vital function of the blood, the transport of oxygen to the tissues in two ways: one is the displacement of oxygen; the other is that the presence of COHb makes the release of the oxygen from the haemoglobin more difficult. It appears that COHb levels greater than 4% are potentially hazardous since at this level psychomotor changes can be detected and there is an increased risk for persons with pre-existent cardiovascular disease. Because carbon monoxide levels tend to fluctuate with traffic conditions, relatively short-period monitoring is indicated.

In some areas, particularly where there is heavy automobile traffic and strong sunlight, photochemical or oxidant pollution is a problem. These reactions involve the hydrocarbons and oxides of nitrogen emitted from automobiles and result in the production (in addition to these compounds) of ozone, various aldehydes and nitrated olefinic hydrocarbons, peroxyacylnitrates and other complex compounds. These can produce significant eye, nose and throat irritation, change in pulmonary function and precipitate asthmatic attacks. Ozone and nitrogen dioxide have been shown in laboratory experiments to enhance the pathogenicity of certain bacteria, to alter structural material of the lung and to interfere with the normal function of certain pulmonary cells. Because of all of these effects, it would seem prudent to monitor for these compounds in the areas where they are likely to occur. It may well be that as our knowledge increases, we should include other substances, particularly in the vicinity of certain industrial operations that give rise to specific emissions. Some of the substances that might deserve this type of monitoring include lead, cadmium, mercury, beryllium, asbestos, fluorides, arsenic, nickel, manganese and polycyclic aromatic hydrocarbons.

ANALYTICAL METHODS

Following the publication by WHO in 1969 of a Guide to the Selection of Methods for the Measurement of Air Pollutants [1] , it was decided to prepare a set of laboratory monographs on the sampling and measurement of common air pollutants. These were selected on the basis of the criteria that (i) where possible, the methods were to be well-tried and their number restricted to facilitate the selection of the most suitable one in any given situation, and (ii) where possible, the methods were those requiring simple equipment; if the only methods available called for complicated apparatus then such apparatus should be readily available in most laboratories.

In consultation with a number of experts and our International Reference Centres, a number of individual monographs have been prepared on the routine sampling and measurement of the common air pollutants sulfur dioxide, suspended particulate matter, carbon monoxide, photochemical oxidants and oxides of nitrogen. The selected methods described in the monographs provide those carrying out surveys with a number of reliable methods from which they can choose one suited to their needs, their expertise and their funds. Where possible, the methods are well-tried and few in number to facilitate the selection of the most suitable in any instance. Also, where possible, the methods adopted are those requiring simple equipment which is available in most laboratories.

Comparison methods, one for each pollutant, have also been decided upon. The comparison methods will allow the results determined by the different routine methods to be evaluated on a common basis. Such a method is not necessarily the most sensitive or the most accurate; it should preferably be well-tried, require simple equipment, the limitations and restrictions in using it in the field must be known and it should be subject to a dynamic or a primary calibration procedure.

A group of experts, drawn largely from the WHO reference centres and collaborating laboratories which met at the IRC on Air Pollution Control, North Carolina, USA, in November 1972, considered and finalized the drafts of the selected routine methods and comparison methods. It is intended that the monographs for each of the common pollutants referred to above will be consolidated into one manual for publication by WHO in the early future.

PROCESSING OF AIR QUALITY DATA

For some time now WHO, in collaboration with the International Reference Centre for Air Pollution Control, has been looking into methods for the handling and statistical analysis of data obtained from the WHO International Reference System. During an initial pilot study, a small number of monitoring sites only will be used (at the present time this involves one urban conurbation in sixteen countries, each of which has three sites, one each in the categories of "inner city-commercial", "inner city-industrial" and "surban-residential") and the pollutants which are monitored are restricted to sulfur dioxide and suspended particulate matter, to ensure that areas where operating problems are shown to exist can be identified and such problems eliminated prior to fully implementing the monitoring programme. The system will then be expanded to include the full complement of monitoring sites and the substances monitored will later be expanded to include other common air pollutants such as, for example, sulfuric acid aerosol, oxides of nitrogen, carbon monoxide and oxidants.

Statistical analysis will involve, for example, arithmetic and geometric means and their respective standard deviations for various time intervals, as well as frequency distributions giving the percentage of sampling intervals during which any given pollutant concentration is exceeded. Facilities for more elaborate statistical analysis for research studies or other special uses will also be available.

REFERENCES

1 M.KATZ, *Measurement of Air Pollution: Guide to the Selection of Methods*, WHO, Geneva, 1969.

SOME RECENT TRENDS IN ENVIRONMENTAL POLLUTION CONTROL AND SOME CURRENT THOUGHTS

GRAHAM CLEARY

Environmental Pollution Division of Environmental Health, World Health Organization, Geneva, Switzerland

INTRODUCTION

The way in which life evolved has been profoundly influenced by the constraints imposed by the biosphere and the products of the evolutionary process in turn modified the biosphere.

The position of man is unique in that he has deliberately altered these processes. While the human population was small and man's productive activities were also small, the environment was large enough to assimilate his waste products. Later, with the industrial revolution throughout the old world, economic motives, coupled with ignorance or apathy, tended towards the maximum use of disposal facilities in the air, in water bodies and in other nearby and available resources. The enormous increases in population, in industry and technology which have occurred in more recent times, especially during the present century, have led, in many instances, to an overtaxing of these resources at a rate which could hardly have been foreseen even fifty years ago.

Historical events have led to a concentration of communites of people in certain defined areas the size and location of which, in a number of instances, bear no logical relationship to the present population's density or needs.

DETERMINANTS OF THE POLLUTION PROBLEM

In addition to the prevailing natural conditions, three factors in particular determine the magnitude and nature of the pollution problem. These are: (i) the degree of industrialization and of economic development generally; (ii) the level and use of technology and (iii) the size of human populations.

Although it is true that a large factor in the pollution of industrialized countries results from the unbridled application of technology, the same problem and the same pollutants are encountered in developing countries as they become increasingly industralized. At the same time it is important to realize that the stage of development affects the perception of pollution as a problem and therefore the resources and priority assigned to dealing with it.

While air pollution generally is considered to be more of a problem in the more affluent and industrialized communities, this is not invariably the case. For example, the burning of cow-dung in domestic fires has caused heavy smoke pollution in some cities in India. Heavy smoke pollution from domestic fires has even been shown to be a problem among native people in remote areas in the highlands of New Guinea [1].

ENVIRONMENTAL HEALTH PROGRAMMES

It is now abundantly clear that at least some of the man-made changes in the biosphere have had a significant impact on human health and in enlightened public health practice today , a good deal of emphasis is placed on the need for the proper management of natural environmental resources. This has been brought about to a large extent by the impact of economic considerations in the growth of many under-developed countries. For many millions of people in these ares, there is a plethora of poverty, hunger and ill-health and these are inexorably linked. It is axiomatic therefore that, for a public health programma to have any chance for success in these parts of the world, concurrent programmes to alleviate poverty and hunger must also be undertaken.

The migration of rural dwellers to urban areas, which became evident in the eighteenth and nineteenth Centuries, has become much more pronounced this century. As a consequence, physical congestion, poor community services, pollution of the immediate human environment by refuse and waste products, squalor and disease, have become all too familiar.

ENVIRONMENTAL PLANNING

The failure to develop comprehensive environmental programmes often in the past has been due to short-sightedness and social negligence, but also to the fact that problems involving these dimensions do not fit into the scheme of the analytical methods of the traditional sciences. It is only in recent times that the concept of an interdisciplinary approach to the problem has become more than an academic ideal.

It is extremely difficult to carry out rational environmental planning by acting under the pressure of emergency as is now often the case. Moreover, in planning for the future, it is necessary to consider that man will continually bring about evolutionary changes through his inherent creative abilities: the concept of an optimum environment is therefore unrealistic because it implies a static rather than dynamic man-environment inter-relationship.

CONTROL OF AIR POLLUTANTS

There are, broadly, three contrasting systems for the disposal of atmospheric pollutants, namely, containment, dispersion and replacement of technological processes by new ones which produce less air pollution.

Increasingly regirous controls have been applied to emissions from industrial processes in recent years. By and large, the technology for control is known but in some instances, such as the control of sulfur dioxide and nitrogen oxides from power plants and other large industrial sources, work still has to be done to make the control economically viable.

In the case of gasoline-powered motor vehicles, a number of countries in the world now have legislation limiting at least some of the pollutants. This legislation will undoubtedly become much stricter in the future and involve much finer control of ignition and carburettion and the use of thermal or catalytic afterburners and oxides of nitrogen by-pass converters.

In the longer term, it is likely that alternative transport systems will be developed. These may take any one of a number of paths. For example, the development of the gas turbine or other power sources, e.g. LPG, steam, fuel cells, etc. It is also possible that the development of highly efficient inter-urban transport systems might obviate the necessity for cars in inner city areas.

Better planning of road transport can improve the air pollution situation in many congested inner city suburbs, but it does not necessarily reduce in total emission of pollutants to the atmosphere.

Dispersion relies on atmospheric diffusion phenomena to dilute the contaminant to such an extent that ground level concentrations are innocuous. The reason for using dispersion in modern plant is usually economic, namely that technology has not advanced to the stage that the pollutant in the volume emitted can be controlled economically (the SO_2 emissions from power stations are an example of this). However, in the future it is likely that dispersion will progressively become a less and less acceptable form of "control".

Replacement of a technological process that causes air pollution by a new one that does not is not always easy to apply in practice. The problem is that the new process has not only to be technologically equivalent to the old one in all essentials (such as the quality of the final product) but it also has to be satisfactory as regards the cost of production. In practice these aspects are not very often compatible.

Location of industrial plant

Unfortunately, air pollution considerations have, in the past, received scant consideration in planning, either on regional or local government levels in many parts of the world. Indiscriminate siting of dwellings and factories discharging pollutants has produced conditions under which the people of certain areas are exposed to potentially harmful concentrations or have serious interference to their well-being and amenity which could have been avoided.

Factors related to environmental pollution have acquired considerable importance in recent years. Factory management has been forced to a broader view in site selection by the realization, whether forced by law or by other pressures, that such selection should not have a degrading effect upon the community. From an air pollution viewpoint, site selection should take into consideration the nature of air contaminants, the efficiency of available control devices, pertinent meteorological factors such as the climate and meteorology of the locations under consideration and the dispersive ability of the air at each possible site, the present quality of the environment at that site and the potential effects on the surrounding areas.

It has come to be recognized that good practice now consists of physically separating people from polluting industries. This applies especially to air pollution (and noise) and is usually accomplished by the provision of suitable buffer zones of either parklands or of light non-polluting industries. The extent of the buffer zone depends upon the type and magnitude of the industry concerned and its potential to pollute.

WHO INVOLVEMENTS IN AIR POLLUTION CONTROL

The ultimate aim of the WHO programme in air pollution is to assist Member Countries in controlling pollution. The Organization is therefore vitally concerned, not only in studies of the health effects of air pollution (human health being considered in its widest sense, as defined in the Constitution of the World Health Organization, namely"... a state of complete physical, mental and social well-being and not merely the absence of disease and infirmity"), but also in the methodology of sampling and measurement, and in control, administrative and legal aspects.

The main objectives in WHO's environmental pollution programme are:

(i) The establishment of criteria, levels of exposure and standards to ensure the protection of man. This was recognized in WHA Resolution 24.47 which endorsed the Director-General's proposal for a long-term programme in environmental health and stressed *inter alia* the need to establish and promote international agreement on criteria, guides and codes of practice with respect to known environmental influences on health.

(ii) The design, development and operation of practical air pollution monitoring schemes. These include such aspects as the promotion of uniform and comparable methods of measurement, the introduction of reliable and effective procedures for the calibration of routine sampling and analytical methods and the development of acceptable procedures for recording, handling and statistically analysing data.

(iii) The prevention and control of pollution. The latter involves such matters as an understanding of the phenomena of generation and distribution of pollutants and of their natural degradation in air, water and soil. Although the control of environmental pollution is fundamentally a technological problem (and the provision of guidelines for control and environmental quality management should be of considerable value to member countries, particularly to developing countries) there are other important aspects of pollution control which involve town planning and transportation policies, legal action and administrative procedures.

REFERENCES

1 G.J.CLEARY and C.R.B.BLACKBURN, *Arch.environ.Health, 17 (1968) 785.*

DISCUSSION

SMEETS

I should like to know if the WHO collaborative monitoring network has been established from the basis of existing sampling and measuring stations or did it start from zero? In what.psychological and financial way has this collaboration progressed?

CLEARY

New sites were set up rather than the application of existing ones. The criteria used were shown in the site identification form and the various categories of "inner city-commercial", "inner city-industrial" and "suburban-residential" sites are described.If it was possible to use existing sites we used them, but where it was not possible we did not want to have this second-rate adaptation of existing sites. So mostly they are new sites.

BRASSER

WHO and EPA set up this network. I am sure that EPA and WHO know exactly what they are going to do with the data. But the details of the data are also available to other users. Is there a profile of these other users? Because this profile greatly determines the way EPA and WHO will use their data.

CLEARY

This is a matter for quite serious discussions between the WHO - EPA and various laboratories involved. The problem comes down to how confidential are these data, and how widely will they become available. I think that it is fair to say that WHO wishes to have them as widely distributed as possible. But I can imagine that this may have political consequences, somewhere.

BERLIN

Dr.Cleary indicated that apparently most of the sites had been set up at a later time, let us say at the request of WHO. To what extent are the techniques used in these sites comparable?

CLEARY

It is our intention, when the recommendations in the manual are published, to exert influence wherever possible to help people use these methods. Our concern is to look towards uniform methods of measurement. This does not mean one method of measurement. This does not mean one method of measurement.

STEWART

I would like to ask Dr.Ruff if he has restricted himself to attempting to predict the exceeding of the 0,1 ppm Californian level only, or whether he is trying to predict the peak levels that might be reached subsequently in any day. Does his approach lend itself to more refined prediction than merely the crossing over the single threshold level and how succesfully can this be done?

RUFF

I have not attempted to go further than the primary output that is presented in the paper. It is not a limitation of the technique, it can be extended to include any number of discrete categories. The success of the prediction is largely a function of the data we have to work with and in this case I do not feel the quality of the data is good enough to justify going any further. Now, a few years after these particular data have been collected, the quality of the data has improved to a point where we can start to attempt these things.

STEWART

You did leave out the figures for NO_2 in a more refined prediction. It would seem to me that his would be a key pollutant to include for predicting ozone levels, because NO_2 and hydrocarbons particularly would be the most relevant pollutants; perhaps NO might be relevant too.

RUFF

For hydrocarbons and nitric oxide, data were available during this time. But because of frequent instrument failures I was forced to eliminate the data. As far as the NO_2 is concerned I do agree with you that this is a fundamental primary pollutant in the formulation of ozone, however, it could well be that CO itself is a better measurement of the automobile input to the ozone formulation than is NO_2. There is more noise possibility on the NO_2 measurement to cause problems in accuracy.

JOST

There seems to be an extremely strong correlation between meteorological data and ozone concentrations. In Germany however, we do not have such a strong correlation. The prediction accuracy of more than 90% is even more than the accuracy of our weather forecast. This seems to be due to the different climatic situation so in this case it is not possible to transform experiences from one location to another.

RUFF

I agree that there are differences between regions with respect to ozone formulation; however, I would like to emphasize that this is an extremely small data set, but it is illustrating that is does work and it can predict. However, we should be a little bit hesitant to draw too many general conclusions from this small data set. I would like to apply it to a more comprehensive data base.

THE PROGRAMME ON AIR POLLUTION CONTROL OF THE WORLD HEALTH ORGANIZATION REGIONAL OFFICE FOR EUROPE

MICHAEL J.SUESS

Regional Officer for Environmental Health, WHO Regional Office for Europe, Copenhagen, Denmark

In its efforts to respond to the growing need of and demand for more and more sophisticated air pollution control programmes in Europe, the WHO Regional Office for Europe in Copenhagen has developed, as part of its Long-Term Programme on Environmental Pollution Control, also a long-term programme for the sector on air pollution control (see Fig. 1). A second group of activities on this subject has been generated through the United Nations Development Programme (UNDP) financed country projects for which WHO acts as the executing agency. Moreover, the long-term effects of air pollution on human health are being studied through pilot and field studies in a number of the European Member States.

The establishment of an automated monitoring system by itself does not solve any problems and should be developed only when necessary for serving well defined objectives. Therefore of not lesser importance, and in a way perhaps even of more importance, is the development of the overall air pollution control programme, including the methodology for the data analysis and its interpretation, as well as the development of appropriate mathematical models for the study and prediction of air quality conditions, with all the consequences involved. Also, not to be forgotten is the need for the training of the appropriate technical personnel for the development and maintenance of the control programme.

Toward this end the Regional Office for Europe has prepared a Manual on Air Quality Management (which is expected to be published later this year) and has started the preparation of a second Manual on Industrial Air Pollutants. The Manual on Air Quality Management includes the following chapters:

- Principles and procedures for the design and operation of air quality surveillance systems.
- Guidelines for preparation of an air pollution source inventory.
- The definition of ambient air quality standards and their application in air management programmes.

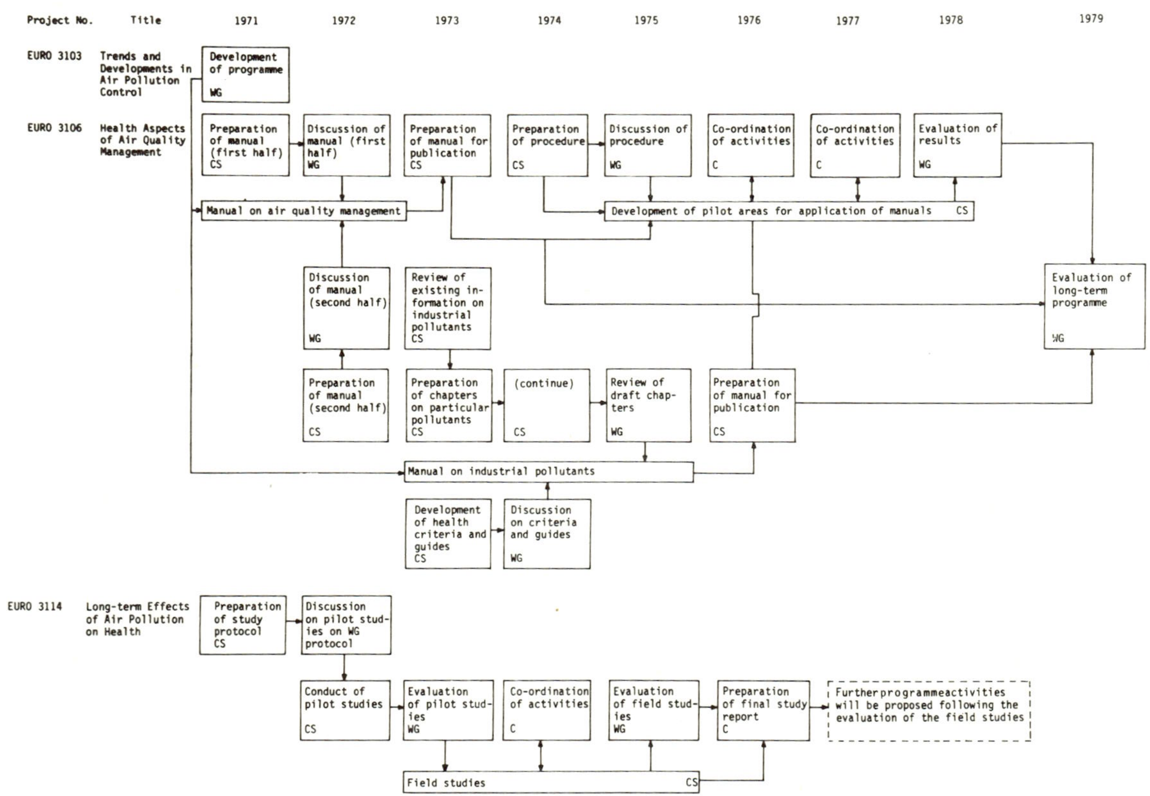

Fig. 1 - The long-term programme on air pollution control of the WHO regional office for Europe

- Legal and administrative aspects and manpower requirements.
- Air pollution meteorology.
- Economic aspects of air pollution abatement.
- Town and country planning.
- Review of existing national regulations referring to criteria and standards.

These chapters were drafted by various specialists and reviewed and discussed by two Working Groups convened in Düsseldorf and Frankfurt, Federal Republic of Germany, in 1972. The Manual also includes a section on Air Quality Criteria and Guides which has resulted from the work of a WHO Expert Committee meeting in Geneva last year. While a chapter on automated monitoring and data analysis has not been initially included, it could now be developed on the basis of the material presented to this symposium and provide a useful addition.

To ensure the Manual a maximum impact at all levels of governments' administration the Manual will comprise of two parts. The first part, addressed to senior administrators (i.e. decision makers and appropriators of funds) should be a short document consisting of well-integrated summaries of the mentioned chapters, giving the points which must be taken into consideration when promoting an air pollution control programme, namely, principles, available methodology, approaches, and possible alternatives. The second part, the detailed chapters themselves, will give the background and detailed information on the respective subject, references, present practice, available criteria and legislation, etc. While this Manual has been prepared by the Regional Office for use in Europe it could equally be adapted, with whatever changes and/or amendments necessary, for use by other countries outside of the European region.

With respect to the second Manual on Industrial Air Pollutants it should be mentioned that it will be compiled on the basis of pollutants rather than on the basis of polluting industries, and that criteria and guides for these pollutants will be developed.

The existence of an air pollution control component as part of the Regional Office's comprehensive environmental pollution control projects in Czechoslovakia, Greece, Poland and Romania enables the assistance to these countries in the assessment of their air pollution problems and the development of control programmes for the improvement of the present air quality, possibly also using automated monitoring networks for that purpose.

Going one step further, a project is under way to study and examine the interrelation between the chronic respiratory diseases in children and the air pollution prevailing in their area of residence. This study, which has started last year, could very well benefit from a continuous, reliable and comparable air quality recording and reporting, for which existing automated monitoring systems and data analysis would be very useful. This cannot be over emphasized with the present experience indicating that data collected from various countries are not easily correlated because of the variety of monitoring programmes and techniques as well as analysis methods which are at present employed.

Finally, it should be emphasized that the Regional Office for Europe collaborates very closely with its Headquarters in Geneva in the promotion and implementation of the overall WHO programme on air pollution control. The Regional Office also keeps in continuous contact with other international organizations, such as the Commission of European Communities (CEC), the Economic Commission for Europe (ECE), the Organization for Economic Cooperation and Development (OECD), the World Meteorological Organization (WMO), etc. in order to exchange information on programmes, complement each other's work and avoid unnecessary duplication of efforts.

AUTOMATIC AIR POLLUTION MONITORING IN THE UNITED KINGDOM

H.N.M.STEWART

Warren Spring Iaboratory, Stevenage, Herts, United Kingdom

The Warren Spring Laboratory is a central government laboratory in which the Air Pollution Division is responsible for co-ordination of national measurement of smoke and sulphur dioxide. In addition the laboratory carries out measurement of pollution of ambient air from specific types of sources e.g. traffic, metal industries etc. The measurements are processed and interpreted for the purpose of the guidance of central and local government on the levels which are arising and on the significance of these levels in the light of guidelines or criteria provided by health autorities and others competent to do so.

In the early 1950's there was a serious incident of air pollution as a result of the build up of smoke and SO_2 concentrations during prolonged inversion conditions. In 1956 the emission of dark smoke by industry was prohibited by the Clean Air Act and in addition Local Authorities were given the power to prohibit burning of all but smokeless fuels in domestic heating systems in areas designated at their discretion.

The National Survey of Smoke and SO_2 took its present form in 1959. From that date the smoke and SO_2 measurements being made by local authorities and others were expanded and there was a standardisation of methods. The monthly seasonal and annual mean values are published for each site annually by Warren Spring Laboratory as well as the number of days exceeding various levels of pollution.

There are about 1200 sites in all, but only about 900 produce valid data in any one year. The survey measures SO_2 as net strong acidity and smoke as dark material in the air by its soiling effect on a filter. These measurements have formed the basis of most of the epidemiological studies on the health effects of smoke and SO_2 and also provide a body of long term and continuous data showing local and national trends since the Clean Air Act became law.

The survey is not automatic except for the use of a sampler capable of taking 8 x 24 hour samples in sequence at some sites to reduce the supervision required.

The analysis is performed and the data are recorded manually locally and partially processed before transmission to Warren Spring Laboratory for final computerised processing.

There is only one fully automatic SO_2 monitoring station being operated by the laboratory. This is located in Central London and continuously records SO_2 concentrations by a number of techniques including those based on conductivity, coulometry and flame photometry. The instruments can be interrogated by telephone from Stevenage, some 35 miles distant. There is also an automatic telemetering warning system which transmits a signal to Warren Spring Laboratory whenever a present level is exceeded.

This site serves the dual purpose of continuous monitoring for central London and also of providing an opportunity for evaluating sensors which could be used to extend the system if this were thought necessary and desirable.

Apart from this site, the laboratory operates a number of sites as part of a traffic pollution study. These are semi-automatic in that data from a number of sensors for gaseous pollutants is recorded on punched paper tape at 5 minute intervals. The pollutants include CO, NO, NO_2, O_3, SO_2 and total hydrocarbons. All are measured by gas-phase physico-chemical methods. In addition meteorological data such as wind speed, direction, radiation intensity and visibility are also logged automatically. Calibration checks are made manually at a minimum interval of a week. Tapes are processed off-line by computer at Warren Spring Laboratory with programmes of data management as required. In addition, samples of particulate matter are taken automatically in a 3 hour sequence for subsequent analysis for lead and for dark smoke. No advantage is seen in on-line real-time data handling. A useful function for direct interrogation of each site by telephone link would be to check on functionality but local authority cooperation means that the sites are visited regularly. A total of 6 such sites is involved in this programme. Some have yet to start operating.

The Laboratory is also conducting a survey for a modelling study in the Forth Valley area of central Scotland. There 30 sites sample SO_2 and smoke automatically over 24 hours periods but the analysis is done manually. Some sites are equipped with automatic directional systems which bring them into operation when the wind is blowing in a preselected sector.

In addition the Central Electricity Generating Board have carried out studies on the behaviour of the plumes from tall stacks by deployment of continuous SO_2 sampling systems around power stations. These have included on-line real-time data collection and processing and micro-meteorological measurements have also been included. Other specialised studies have included sampling systems which operate automatically and we have carried out several experiments with rapid data transmission employing multiplexing over long distances for demonstration purposes, e.g. from Tokyo to Stevenage, on public service telephone lines. Some work on local radio transmission of measurements from sensors attached to balloon cables has also been carried out.

In conclusion I would say that we have an interest in obtaining a full appreciation of the possible benefits of on-line data transmission and real-time presentation. We are very impressed with the amount of effort and expenditure which has gone into such systems in other countries. We do not have such highly concentrated groups of refineries in the UK as for example near Rotterdam. This is relevant. I feel it is worth remembering, although it may seem obvious, that measurements in themselves cannot directly improve air quality. It is often implied that they can. No doubt we do feel better when the physician takes out temperature but it is what action he takes following that measurement that really affects us directly. Perhaps we are in danger of believing something is being done just by making more measurements.

The main emphasis should perhaps be on forecasting of episodic high pollution levels from meteorological data. The exceeding of agreed criteria and subsequent warning represents failure of the system to respond quickly enough. Often, it is the detailed meteorological data rather than the pollution data that are lacking for such predictive forecasts.

The UK Meteorological Office has been developing 24 hour predictive model for the level of sulphur dioxide pollution in London. This has been produced using meteorological data and the daily sulphur dioxide data from four central London sites. It seems to us that a model operating on meteorological forecasts and a minimum of pollution data for 12 or 24 hour periods is a realistic and worthwhile approach to the prediction of high pollution levels in a normal urban area not subject to sudden, significant increases in pollution emissions such as might be produced by a particular industrial plant. A predictive model on this sort of time-scale offers a real possibility of applying control measures in time to avoid the worst pollution levels.

It is now possible to buy hardware for air pollution measurement and for data acquisition and transmission of any desired sophistication. Certainly we must note that maintenance calibration and validation of the data will be more time consuming than we have perhaps anticipated.

I would welcome specific moments from those who have experience with automated networks, particularly where data is processed and presented on-line and in real time on a national basis. The following points are of interest:

1. Where do such systems have particular benefits? At what level of sophistication are they essential?
2. What is the minimum requirement for an adequate network for monitoring purposes and for warning purposes?
3. When does real time data accessibility become indispensible? Is this relevant only for a control or enforcement activity?
4. What reduction in pollution results from the control aspects, as opposed to the measurement aspects, of the total programme?

The viewpoints which can emerge on these points would be of considerable value to those who may be considering the situation and wish to do so in the light of the experience gained by others.

DISCUSSION

BURMANN

Dr.Suess said automatic monitoring systems are available, it is just a matter of money. Did you have anything specific in mind?

SUESS

I do not really have something in mind. But I used this opportunity to introduce the project on chronic respiratory diseases in children which are being studied in relation to air pollution. Children are picked rather than adults because they are not moving as much out of their own zone. It is the easiest group to work with.

CLEARY

Two of the important reasons for choosing children are, first they have no occupational exposure, and second they are hopefully non-smokers.

SUESS

We use children in the agegroup of 8-10 because they have to be old enough to collaborate with the examiners and young enough to be non-smokers. But we found in Denmark that they have added special questions on smoking because some of them seem to smoke already at that age.

WEBER

Dr.Stewart told us about the prediction of air pollution. In the U.K. they try to predict high concentrations and when there is an episode coming they start more intensive measurements. Is there a special technique you use for this purpose. According to my experience there is no way of predicting air pollution without having continuous measurements of some air pollutants. It may not be possible to get these predictions only from meteorological forecasts.

STEWART

Most of the data we have in the U.K. are based on 24 h measurements and so are most of the criteria values we are concerned with. For some pollutants which are generated by traffic, we are concerned with hourly values and we measure these as appropriate. For example we find that for elevated ozone levels, which we are interested to measure, that we can define the meteorological conditions which give rise to these. Our difficulty is not to get sufficient pollution data, which we are measuring on the basis of short term measurements in this case because of the lack of depositive data, but the difficulty in getting the prediction model is that there is not enough detailed map data available to give the synoptic situation over a long period in the past. The sort of question I was seeking to raise is that we may not need more on-line data for any given pollutant, but we may need other kinds of data. We may identify other kinds of data. We think we can predict the times, the situations which will give rise to ozone, but we are not able to test this model because we do not have enough map data to test it properly. For SO_2 it is a little more encouraging in that we only require 24 h values.

RUFF

Dr.Stewart's problem is that he seems to have sources scattered about London that would make an automated network a bit expensive for his purposes. He should consider the possibility of setting up mobile vans for short periods around selected sources. This has been done for similar problems in the U.S.A.

SHEARER

We find it a rather effective way to maximize the use of limited funds with respect to attempted monitoring over great distances.

WEBER

Besides the four traffic measuring stations in London Dr.Stewart mentioned one background station. Where is this located?

STEWART

The short answer is that it is in the city of London. And the purpose of it is to attempt to decide the relative contribution of fixed sources in the city compared to specific traffic pollution. We are using it in conjunction with a traffic-side site.

CLEARY

I have a question in relation to the paper that Dr.Berlin presented at the symposium on lead in the environment in Amsterdam in October last year. Comparability, particularly across different countries, concerns me a lot. Would you mind going over the range of scatter in the results that you got in the analysis of lead in blood?

BERLIN

We did an intercomparison programme about lead in air with about seven sites. Now we have completed that study. We used two "reference methods" plus the local method and obtained rather good results. The scatter was not more than about 15 to 20% for the lead in air. For the lead in blood we had about 26 laboratories to which we sent identical blood samples. The order of magnitude was quite correct; the actual concentrations were around 30 μg and the range of the obtained values was from 10 to 110 μg.

SHEARER

Let me give you just an additional example of experiments that are hard to control: we recently participated in a control experiment on sulfur dioxide analysis in air. 174 laboratories in our country participated in this; we sent to each of these laboratories 5 different concentration ranges of sulphur dioxide, asked them to analyse these samples as they would the routine samples in their laboratories and report the results back to us. We have finished the evaluation of these results and prepared a report. Of the laboratories participating, slightly more than 50% were unable to get the the correct result on all 5 samples and the ability of the laboratories to get the correct results decreased as the concentration range of the samples got higher. At the highest concentration range I think less than 20% of the laboratories got the correct results. We are intending to repeat this and we have underway the same type of experiment for carbon monoxide, for hydrocarbons, and for particulate matter.

SUESS

For the purpose of our study on the chronic respiratory diseases in children in relation to air pollution we have sent a control team around. It was composed of two persons, one an epidemiologist who was to look into the procedures used in examining the children and administering the questionnaire that was involved, and the second person an air pollution expert who was looking to the recording of data, because after all you want to make sure that the air pollution data are equally comparable. We have found out that different countries use different methods for monitoring and for analysing the samples and in some cases there is no correction or correlation possible. This means that something is wrong in monitoring. I like to emphasize that is is extremely important to unify and to standardize so that the data can be used for the study of the effect on man and the ecosystem at large.

SHEARER

I would like to make one comment at this point. I think that to-day an attempt has been made to stress that when you are contemplating designing, implementing and putting into operation a complete monitoring system, you must lay out some careful objectives for what this monitoring system is attempting to obtain. You must give very careful consideration to the various types of individual sensors that you need to put into this system; but you must allow yourselves some flexibility to improve the various components in this system as technology increases.

RUFF

I would like to know from Mr.Nordø how he derived the decay constants for SO_2.

NORDØ

The numerical transport model for SO_2 and H_2SO_4 is based on the continuity equation. From this relation an equation for the decay time can be derived. This equation contains constants, which are adapted in such a way that there is a good agreement between the decay times published in the literature [1,2,3] and the calculated decay times. Of course this approach is just an introduction. We are able to derive lots of equations to describe the transport in more detail; but due to the lack of figures, the results are so uncertain that in my opinion you have to go in the atmosphere and see what is going on there e.g. by aeroplane measurements.

STEWART

What was the inversion height or layer thickness for a concentration of 100 $\mu g/m^3$?

NORDØ

In the Christmas episode it was 700 meters, in the other episode it was 1300 meters. In these cases there was a marked inversion. But this is not always the case.

REFERENCES

1 C.E.JUNGE, *J.Geophys.Res.* 65(1960)227

2 A.R.MEETHAM, *Quart.J.Royal Meteorol.Soc.* 76(1950)359

3 E.WEBER, *J.Geophys.Res.* 75(1970)2909

NATIONAL MONITORING SYSTEMS IN THE U.S.A.*

FRANZ.J.BURMANN

U.S. Environmental Protection Agency, National Environmental Research Center, Research Triangle Park, North Carolina 27711, U.S.A.

To discuss national monitoring systems in the United States, it is necessary to outline briefly the historical development of the U.S.Environmental Protection Agency's (EPA) monitoring stategy. EPA, in August 1971, promulgated guidelines for the States in the preparation, adoption, and submittal of implementation plans. In the guidelines, minimum requirements were specified that must be met by State air quality surveillance systems as part of their implementation plan. After approval of the implementation plan, the States are allowed 2 years to complete their monitoring systems. State and local governments have the primary responsibility for maintaining surveillance of air quality of those pollutants for which standards have been promulgated. To achieve this goal in a timely fashion, EPA is working with state and local agencies to evaluate their present systems and to assist them with technical problems. In addition, program grant funds are provided by EPA to be used by state and local agencies for the establishment and operation of air monitoring networks.

EPA's monitoring activities are carried out in two ways: though networks and special studies. The networks are made up of permanent stations located in urban and nonurban areas throughout the nation. They serve to identify air quality and regional and national trends. Where needed information is not provided by the network, special studies are performed.

One of the oldest, if not the oldest, air monitoring network in the world is the National Air Sampling Network (NASN), which had its beginning in 1953. At that time, the U.S.Public Health Service initiated monitoring stations to measure solid pollutants in 17 cities. State and local agencies supplied the manpower to collect the sample; the Federal government provided all necessary equipment, supplies, maintenance, and analytical services.

As more and more communities wanted to participate, the NASN expanded. By 1957 there were approximately 66 stations in the network. During the following years, public concern for air pollution increased, and public health officials began to wonder about gaseous pollutant levels in the air. Therefore, in 1960 NASN began limited sampling for gaseous pollutants.

*The mention of company or brand names is not to be considered as endorsement or recommendation for use by the Environmental Protection Agency

The NASN continued to increase until it totaled approximately 200 urban and 30 nonurban stations by 1966. Each following year witnessed the growth in the number of participating agencies. By 1972, the NASN operated approximately 250 urban stations to collect particulate and 175 stations for collection of sulfur dioxide and nitrogen dioxide samples. In addition, 25 nonurban stations were maintained for each of the aforementioned pollutants.

The stations are equipped with integrating samplers to obtain 24-hour sulfur dioxide and nitrogen dioxide samples and high-volume samplers to obtain particulate samples. These stations are activated every 12th day. All samples are routinely returned to a central laboratory for analysis.However, with the initiation of monitoring efforts by the States and the emphasis on decentralization of routine monitoring activities at the Federal level, the responsibility for the NASN is gradually shifting to EPA's Regional Offices. EPA also operates networks that monitor radioactive particles, air pollutants dissolved in rainwater, and particulates collected on membrane filters.

Another EPA network, the Continuous Air Monitoring Program (CAMP), has been in existence since 1962 in six major cities throughout the country. The cities were selected to represent as wide a variety as possible of the factors that influence air pollution problems, such as climate, topography, type of industry, principle fuel used, and so on, thereby providing a base of information as broadly applicable to all our urban areas as six stations possibly could. Originally envisioned purely as a research tool, they have been invaluable because of the uninterrupted data generated for 10 years. Carbon monoxide, nitric oxide, nitrogen dioxide, sulfur dioxide, total hydrocarbons, and total oxidants have been measured continuously, and the data have been utilized for setting baselines and determining trends.

In addition to housing the continuous instruments, the facilities served as testing stations for experimental networks and for field evaluation of instrumentation employing new methodology. Since CAMP's initiation, an effort has been made to utilize the best available methodology. However, changing from one to another instrument or from one method to a newer method poses great difficulties in air monitoring. The usual problems are compounded by a constantly changing sample composition, making comparisons extremely difficult. The best approach to assure continuous, valid data under these conditions is parallel sampling of sufficient duration with the instrument under investigation. For example, the continuous instrument used to monitor ambient nitrogen dioxide in the CAMP employs a modified Saltzman methodology. Latest information available indicates that ozone in high concentrations is and interferent in the methodology. Therefore we are presently engaged in gathering nitrogen dioxide data not only by the Saltzman methodology but also by the chemiluminescence method, which has no known interferent.

This approach enables comparison of data gathered by different methodologies sampling diversified ambient pollutant compositions. It affords both the opportunity to evaluate the method and the instrument employing the method under variable field conditions, and also the opportunity to gain necessary expertise in the operation, maintenance, and calibration of such a system. In addition, because of the geographic distribution of the CAMP stations, they can be used as focal points for demonstration of monitoring systems to state and local agencies. They also make ideal sites for field testing of experimental networks and of equipment developed for specific pollutants.

It is only proper to mention the disadvantage of CAMP, which is the site location and the fact that the previously mentioned pollutants are monitored at the same site. However, the site criteria initially employed was sound and still has applicability for this type of station. Most NASN and all CAMP sites were located in a downtown area of a city, where there is rarely space for the construction of additional sources, and where peak traffic flow is near saturation. In other words, as far as ambient air pollution is concerned, the location was the most stable one in the city. It is, of course, acknowledged that measuring air pollution levels only in the relatively stable downtown area of a city does not supply enough information to detect trends in overall pollution levels in a metropolitan area. Furthermore, it is also recognized that if one intends to monitor in areas of maximum pollutant concentrations, sensors for different pollutants should be located in different areas. However, the advantages of having a historic pattern of the major gaseous pollutants for 10 years far outweigh the disadvantages just mentioned.

EPA's Community Health and Environmental Surveillance System (CHESS) is a national program designed to relate ambient air quality to community health. It is anticipated that the program will be operational for toxic substances in other media by 1975. The program is conducted in cooperation with local public health agencies, institutes of higher learning, and private institutions. Its purpose is to evaluate existing standards, to obtain intelligence for new standards, and to document the health benefit of pollution controls.

Presently, the network consists of six geographic areas. Each geographic area comprises a group of three to seven communities selected to represent an exposure gradient for designated pollutants. The possibility of detecting pollutant interactions is enhanced by selecting communities with principal exposure to one gaseous pollutant alone, to particulate alone, and to a combination of the two pollutants.

Initially (in 1969), most stations used manual instruments to monitor 24-hour average sulfur dioxide concentrations and cyclone samplers to assist in collecting respirable particulate matter less than 5 micrometers in size. Since January 1972, the 24 hour sulfur dioxide samplers are being replaced with automated instruments. The updated air surveillance system in the CHESS program is called the Continuous Health Air Monitoring Program, abbreviated CHAMP. It is expected that the CHAMP will include approximately 50 continuous sulfur dioxide and nitrogen dioxide instruments, approximately 5 continuous carbon monoxide instruments, and approximately 24 oxidant instruments. Selected stations have been equipped with telemetering hardware, permitting data processing at a central facility. With this system, it may be possible to relate health indications of acute response to short term variations in pollution concentrations.

In addition to EPA's monitoring systems, there are over 3000 stations in the U.S.A. where some type of air quality measurements are taken. Complexity of the monitoring stations varies considerably. An estimated breakdown of activities that relate directly to pollutants for which "National Ambient Air Quality Standards" (NAAQS), have been set is as follows:

Total suspended particulates - high-volume samplers	3000
Total suspended particulates - tape samplers	500
Sulfur dioxide - 24-hour integrating samplers	650
Sulfur dioxide - continuous analyzers	350
Nitrogen dioxide - continuous analyzers	150
Carbon monoxide - continuous analyzers	200
Oxidant (ozone) - continuous analyzers	200

EPA is also required to obtain information on trends of pollutants for which standards may be set in the future. A great amount of such information comes from the total suspended particulate samples, which are analyzed for many trace metals and other elements. The total suspended particulate samples obtained by NASN high-volume samplers were also utilized to establish a data bank on selected compounds and elements found on individual and composite samples. Presently, our data bank contains data on more than 20 air contaminants obtained from particulate samples.

Special networks may be required to fulfill an immediate unexpected need or to satisfy a short-term requirement. It is not the purpose of this paper to dwell on details; however, it seems appropriate to list some of the major tasks involved in establishing a special study network:

1. Determine how many stations are needed, where to locate them, and ascertain the number and type of instruments to be obtained.
2. Develop performance specifications to procure the continuous monitoring instruments, recorders, calibration gases, and other necessary equipment and supplies.
3. Select sites for fixed and mobile stations, make arrangements for operator training, and arrange for initial installation of instruments in a suitable field facility.
4. Develop installation, calibration, operation, maintenance, and training procedures.
5. Design a quality assurance and data validation program.
6. Train the instrument operator in the procedures to be followed.
7. Provide data processing service to assure that acceptable data are being obtained.

In 1970, a monitoring network for carbon monoxide and ozone had to be implemented on short notice. The objective of the project was to assist States in obtaining ambient concentration data for these pollutants. This need arose because at least a minimum amount of air quality data were required to enable States to develop implementation plans within a specified time.

The requirements for the study to achieve this goal can be briefly stated as follows:

1. Provide, within 3 months, an operative continuous monitoring network for measurement of carbon monoxide and ozone in approximately 30 cities.
2. Collect valid data, without interruption, for 3 months from at least one site in the selected cities.
3. Obtain sampling data for each pollutant that reflect conditions of the more heavily polluted areas in the Air Quality Control Region.
4. Employ averaging times consistent with air quality standards, i.e., 1 hour and 8 hours for carbon monoxide and 1 hour for ozone.

A similar study conducted last year had as its objective the reassessment of nitrogen dioxide air quality data in 42 Air Quality Control Regions. The study was necessitated because the method previously employed possessed inherent deficiencies that may result in overestimated ambient nitrogen dioxide levels. To meet the objective, EPA acquired chemiluminescent nitrogen dioxide analyzers, developed new calibration expertise, selected sites, installed, calibrated and operated the instruments to obtain the desired data. In addition, the routine NASN sampler was utilized to obtain additional integrated samples using a modified reagent system to determine inter-relationship between the methods.

A project presently being initiated involves the determination of sulfur dioxide concentrations in the vicinity of 13 smelters. Since the smelters are located in remote areas, data are lacking on which to base emission limitations. Each plant is monitored by two continuous sulfur dioxide analyzers to determine short-term sulfur dioxide levels and four integrating samplers to obtain 24-hour averages.

As previously mentioned, the plants are located in sparsely populated and remote areas, necessitating installation of mobile shelters. The study is designed to obtain data during a 1-year period.

A review of national monitoring systems is not complete without mentioning the planned Regional Air Pollution Study (RAPS). A network of 25 to 40 completely automated air monitoring stations telemetering data to a central data acquisition facility is planned for St.Louis, Missouri. RAPS is designed to develop enough information on air pollution processes and effects in a single urban region to verify existing air pollution simulation models and to develop better models. EPA is presently in the process of selecting a prime contractor to operate the facility.

In conclusion, I would like to summarize some of the principal reasons for advocating an effective monitoring program. They are:

1. To minimize the risk in decision-making relative to the progress being made toward attainment of standards.
2. To insure the actual attainment of the standards.
3. To confirm conclusions reached by using non-EPA data.

In the final analysis, the effectiveness of efforts to control pollution can be determined only by measuring improvements in ambient air quality.

MONITORING AIR POLLUTANTS ON MESOSCALES (20 KM - 500 KM) AND ON LARGE SCALES (500 KM - 5000 KM)

JACK NORDØ

Norwegian Meteorological Institute, Oslo,
Norwegian Institute for Air Research, Kjeller, Norway

On such scales it is possible to compute diffusion by deterministic equations, providing there is information on the instantaneous wind fields. The winds may also be derived from other meteorological parameters, using the equations of motion and thermodynamics.

The local pollution concentrations will in most places depend very much on the large scale wind speed as well as the large scale static stability (vertical stratification of the air). The large scale cloudiness is also important as radiation processes may alter the static stability of the air, and consequently the local rate of diffusion.

Local wind patterns and local stratifications may be computed from the large scale variables mentioned above. In addition one must also consider the local topography, local heat sources and sinks. But so far rather few meteorologists have studied such motions in great detail, including katabatic winds, landsea breezes, mountain-valley winds, and motions induced by the urban heat island. At present the simplest approach will be to derive statistical relations between the local weather and the large scale flow.

It is possible today, with a fair degree of reliability, to forecast the large scale meteorological parameters for the next 48 hours or more. In this way there exists a possibility to take precautions well ahead of time, and reduce the adverse effects of a coming pollution episode.

A brief account shall now be given of the O.E.C.D.project on long range mass transport of air pollutants, a joint effort by ten European countries. Careful intercalibrations have taken place between the cooperating laboratories, and the results show a high degree of compatability. Approximately fifty stations are measuring daily concentrations of various sulphur compounds. Within a year from now one hopes to have a number of upper air concentrations measured by coordinated flights of 3 - 8 airplanes. The bulk of the mass transport of air pollutants take place at levels well above the ground, and only upper air measurements can determine the correct size of this transport.

The data flow is given in Fig. 1. Upper air data observed every six hours by radiosondes and pilot balloons are used as input. The observations are checked and anlysed in near real time by the computing centre at the Norwegian Meteorological Institute. Winds are predicted for the next 36 hours, and the forecast wind fields are used for estimation of the likely paths (trajectories) of selected air masses. Two kinds of trajectories are being computed twice a day. The first version is the "backward"trajectory in time, i.e. the trajectory arriving at a selected point 36 hours later. The "forward" trajectory starts from a source region at 00 GMT (12 GMT), and its position is then printed every six hours during the following 36 hours. In this way one gets a mean to predict the future positions of large scale plumes, and it is also possible to prepare for more extensive measurements in interesting episodes.

"Diagnostic" trajectories are also computed, using observed winds only. In this way the pollution may be monitored in near real time. The observed or predicted winds may also be used for concentration computations, using the equation of continuity and regional emissions of SO_2. The following equations are much simplified versions expressing the average concentrations between surface and a certain mixing height. The mass transport across the upper boundary is neglected.

$$\underset{(1)}{\frac{\partial q}{\partial t}} = \underset{(2)}{-(k_1 q + k_2 q^2)} + \underset{(3)}{\dot{Q}} - \underset{(4)}{\nabla_h \cdot (q\vec{V}_h)}$$

$$\underset{(5)}{\frac{\partial \mathfrak{q}}{\partial t}} = \underset{(6)}{-ck\mathfrak{q}} + \underset{(7)}{\alpha(k_1 q + k_2 q^2)} - \underset{(8)}{\nabla_h \cdot (\mathfrak{q}\vec{V}_h)}$$

The numbers underneath the two equations refer to the various terms.

Symbols

q = mixing ratio of SO_2 = $\frac{\text{Mass of } SO_2 \text{ in a volume}}{\text{Mass of air in the volume}}$

$\mathfrak{q}$ = mixing ratio of H_2SO_4

$\vec{V}_h$ = the horizontal wind. New values are provided every 6th hour, on the basis of the meteorological observations

k_1, k_2, ck, and α are constants. In the computations we have used values close to the following,

$k_1 = 10^{-6}$ sec^{-1}, $k_2 = 3\ 10^2$ sec^{-1}, ck = 2,5 10^{-6} sec^{-1}, $\alpha = 1$

$\nabla_h = \vec{i}\frac{\partial}{\partial x} + \vec{j}\frac{\partial}{\partial y}$, where x and y are coordinates in a cartesian coordinate system, $\vec{i}$ and $\vec{j}$ are the corresponding unit vectors.

This symbol signifies a two-dimensional differential operator.

If the concentration of SO_2 is $C(t_0)$ when the polluted air leaves the source region and diffusion is neglected, the concentration will be reduced to 50% according to Table 1.

Table 1

$C(t_0)$ in microgrammes per m^3	1	10	100	1000	4000
$T_{\frac{1}{2}}$ in units of 1000 sec.	690	250	38	4	1
	(8 days)	(3 days)	10h)	(1h)	(15mn)

(1) and (5) signify the change with time in the mixing ratio of SO_2 and H_2SO_4 respectively, at a fixed point.
(2) is the decay of SO_2. The term makes the decay relatively faster when the mixing ratio is high. The published literature on the subject points in this direction, see next chapter.
(3) is the emission of SO_2. This source function is known to a first approximation from national surveys.
(4) and (8) are the transport of SO_2 and H_2SO_4 respectively, by horizontal winds.
(6) is a decay term for H_2SO_4.
(7) is a source of H_2SO_4. This term makes some of the SO_2 a source of H_2SO_4. The efficiency of this source may be varied by means of the constant α.
For numerical treatment by a computer the equations are re-written in a finite-difference form.

As an application of such numerical equations one may show computations of mass transport of air pollutants across international boundaries. The case studies were done by Mr.Saltbones from the Norwegian Institute for Air Research, and they were considered as preparatory work for the O.E.C.D.-project. Grønskei from the same institute has applied a similar model to the problem of dispersion of pollutants in the urban region of Oslo.

The long range transport of air pollutants has been computed for various values of k_1 and K_2. K_2 has been given values close to 250 sec^{-1}, but k_1 has been varied from 10^{-5} sec^{-1} to $10^{-6}sec^{-1}$. The following illustrations are based on $k_1 = 10^6sec^{-1}$. The episode was predicted 2 days in advance, and the Norwegian Institute for Air Research carried out aircraft measurements on the peak day. The observations were in good agreement with the computed concentrations.

Judging actual records of pollution concentrations it is often hard to decide whether they are caused by local or distant sources. In model studies, however, it is possible to omit nearby sources. In this way one can estimate the pollution level due to long range transport only. For the following episode estimates were computed with and without Scandinavian emission, but the differences were rather negligible in this extreme case. The episode dates back to March 1972 (three months before the O.E.C.D.project was initiated) and it persisted for several days. A relatively dense network of stations provided daily records from Southern Norway, and the verifications are therefore concentrated to this region. The southeasterly flow caused heavy orographic snowfall over Southeastern Norway, and scavenging caused a high deposition rate of pollutants over the countryside with no significant local sources. Using the observed precipitation amounts, as well as measured and computed concentrations, a total deposition of 2000 - 5000 tons of sulphuric acid may have occurred during these 24 hours. The model calculations suggest strongly that this deposition is mainly due to scavenging of pollutants originating outside of Scandinavia.

The presented model calculations are all based on Eulerian integration schemes which has been in use since July 1972. Recently another model approach was completed. It is based on a Lagrangian integration scheme, and it seems to give realistic estimates of long range transport of air pollutants at a low computer cost.

A statistical approach is also in progress. It is based on calculations of "backwards" trajectories of air parcels passing a given O.E.C.D.station every six hours. Whenever a trajectory crosses a grid square a weight proportional to the emission within the given square is added. To each trajectory two weights are assigned, one due to the strength of nearby sources, another due to the size of distant sources. Correlations with measured pollution concentrations may then yield interesting results according to some preliminary studies recently undertaken.

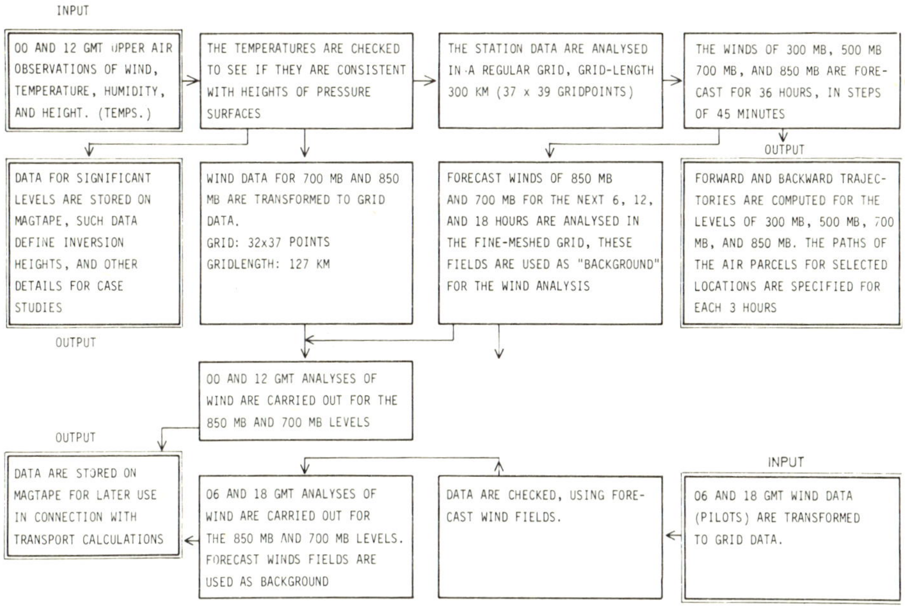

Fig. 1 - Data processing procedures for the O.E.C.D.-project

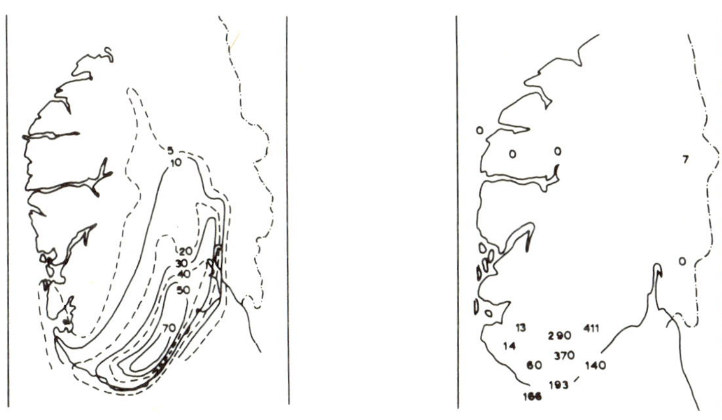

Fig. 2 - Left: precipitation in Southern Norway (mm) during the 24 hours from 07 GMT March 4 to 07 GMT March 5.

Right: precipitated sulphate in mg mer m^2 as measured by the NILU network during the same 24 hours

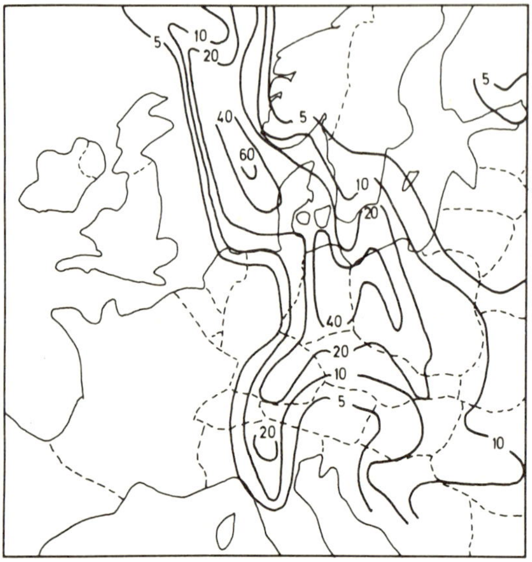

Fig. 3a - Computed SO_4 mixing ratios (10^{-9}) at 00 Z March 4, 1972

Fig. 3b - Computed SO_2 mixing ratios (10^{-9}) at 00 Z March 4, 1972.

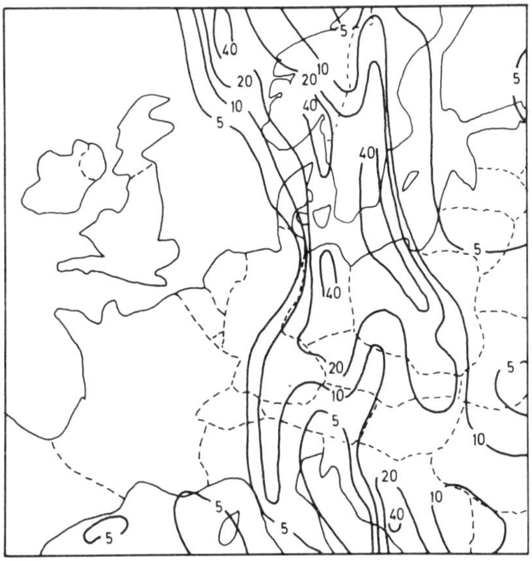

Fig. 4a - Computed SO_4 mixing ratios (10^{-9}) at 12 Z
March 4, 1972

Fig. 4b - Computed SO_2 mixing ratios (10^{-9}) at 12 Z
March 4, 1972

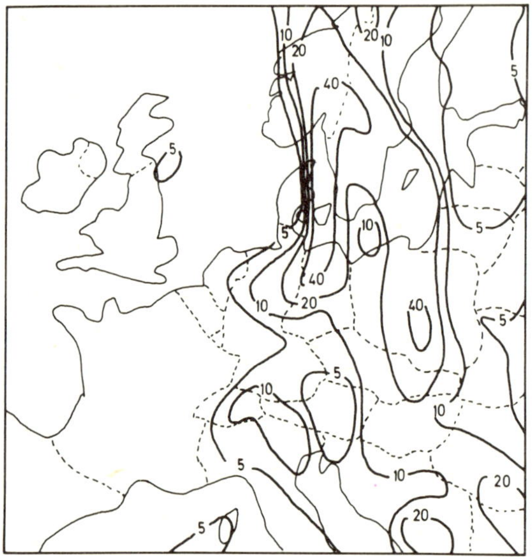

Fig. 5a - Computed SO_4 mixing ratios (10^{-9}) at 00 Z
March 5, 1972

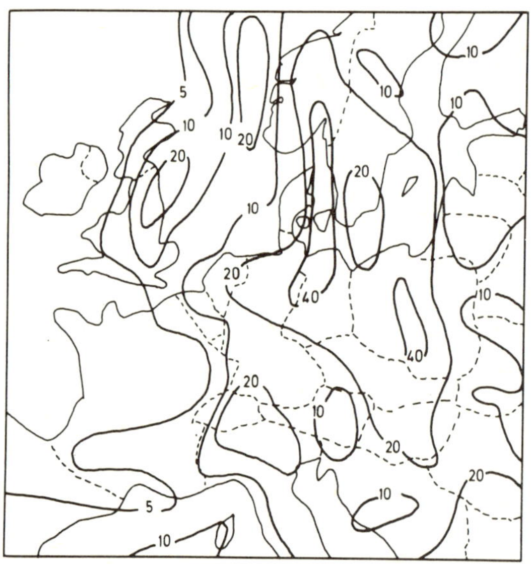

Fig. 5b - Computed SO_2 mixing ratios (10^{-9}) at 00 Z
March 5, 1972

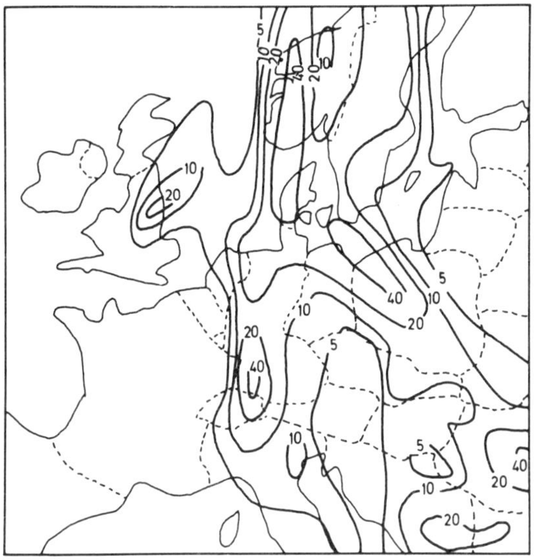

Fig. 6a - Computed SO_4 mixing ratios (10^{-9}) at 12 Z March 5, 1972

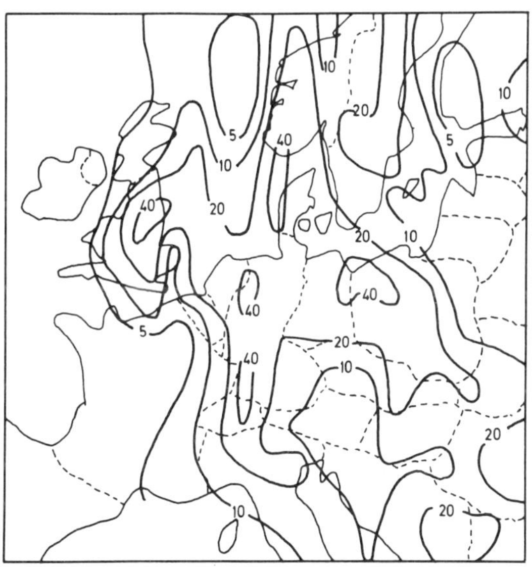

Fig. 6b - Computed SO_2 mixing ratios (10^{-9}) at 12 Z March 5, 1972

A PILOT NET AS SCALE MODEL AND RESEARCH UNIT FOR LARGE AUTOMATED AIR QUALITY MONITORING NETWORKS

J.G.KRETZSCHMAR, G.FIEUW

Departement Meting en Kontrole Stralingen S.C.K./C.E.N., Mol, Belgium

INTRODUCTION

During the last years quite a lot of countries all over the world became aware of the worsening of the quality of their environment, and in the relatively near future everybody will have to face the same problems in the field of air, water and soil pollution.

Although air pollution is may be not the most striking environmental problem, as long as smog and odour problems do not appear, it remains a fact that it is not the easiest one to solve.

As well as protection the monitoring and improvement of the quality of the air we all live in, will create a wide spectrum of problems with an impact on science and technology, industry, economy and the entire society where welfare has to improve well-being.

As nuclear centers had to cope with radio-active pollution; and more specifically with the dispersion, dilution and transport of radio-active material in the atmosphere, long before society even noticed the global pollution threat, the Belgian Research Center for Nuclear Energy (S.C.K.-C.E.N., Mol) felt called upon to tackle pollution problems. In this context the Center started several programmes in the field of air pollution control and abatement with a scientific knowledge and background, characteristic of a nuclear research organisation.

LARGE AUTOMATED AIR QUALITY MONITORING NETWORKS

The continuous control of the air quality in a vast area is only possible by means of a large automated monitoring network where information on the actual situation of the immission, and eventually the emission, is available on-line. This information, together with well-defined and justified air-quality standards from the base for the safeguarding of public health and the protection of crops, vegetation and foods in general.

Although these characteristics and possibilities look quite promising, it must must be emphasized that, no matter how large or how sophisticated an automated monitoring network is, it remains a system or an instrument to pursue many different ends.

The wide range of possibilities does not facilitate the proper choice or the optimum design of the system, while the financial implications are important as well for the installation as for the operation and updating of the network.

A nationwide monitoring network is a rather unwieldy scientific instrument as it is pretty large, as it must run almost continuously to reach certain basic objectives while in addition the procedure for data acquisition, processing and analysis, as the monitoring apparatus, the teleprocessing system and the software subroutines are not easily changeable without causing much trouble and many problems.

The fact that the components of the network are scattered all over the country does not improve the flexibility and tractability of the system.

Despite all those factors that make the instrument rather static, we must try hard to force it to the rather dynamic side. Indeed, the change in the nature of the pollutants, the variation in the emission pattern in time as well as in space, the improvement in the available monitoring, teleprocessing and computer equipment and the evolution, and hopefully,improvement in our knowledge and understanding of the phenomena of air pollution will influence the systems in operation at one time or another.

Taking into account the urgent nature of the pollution problems and aware of the abovementioned difficulties, the S.C.K./C.E.N. decided to start a programme with different phases leading to a final proposal for a fully automated air quality monitoring network. In the first phase a paper study was made of the available monitors and the possible teleprocessing systems together with the development of mathematical models and the necessary computer subroutines to evaluate the actual situation, to predict the trends for the future and to work out a warning strategy and emergency procedure for episodes of high pollution. As a logical consequence of this study, a second phase was started immediately. Besides some exploratory compaigns for SO_2 and particulates and besides analysis of the results from previous campaigns or from semi-automatic networks operating in Belgium, the construction of a pilot net was started in Mol, in order to acquire the required know-how.

OBJECTIVES OF THE S.C.K./C.E.N. PILOT NET

The objectives of the pilot net under construction at the S.C.K./C.E.N. in Mol can be summarized in the following way:

1. Investigations of the different components of an automatical network. Those components consist of the instruments for measuring the different meteorological parameters influencing the air-pollution, the monitors for determining the concentrations of several pollutants, the electronic devices for obtaining and transforming the signals and the communication links and the computer hardware.
2. Development and testing of the required software for the on-line data acquisition, validation, reduction and analysis by means of a mini computer (PDP-8/E).
3. Development of mathematical models and other software for the central computer (IBM 360) in order to evaluate and supervise the actual situation to determine the trends and to obtain the required statistics.
4. The construction of a flexible scale model of a large automated air quality monitoring network using the components and software described in [1] [2] [3] This scale model would allow dynamic research on the behaviour of the complete system under different conditions, normal as well as extreme, in order to determine the feasibility and reliability of a proposed network.
5. The simulation and solution of specific problems that will arise once the real network is operating. As it would be expensive and risky to experiment with the large monitoring network, the pilot net will be a handy instrument to solve the problems of the former.
6. The investigation and evaluation of the newest instruments, techniques and ideas in order to improve the large scale network when the latter has been installed and must run in a continuous way.

This list of objectives is partly tentative as the dynamic character of the chosen approach implies changes in objectives and priorities whenever needed.

PRACTICAL REALIZATION OF THE OBJECTIVES OF THE PILOT NET

As knowledge of the atmospheric condition is indispensable for air pollution studies the *meteorological parameters* must be measured. This is done by means of several instruments installed along a 120 m high meteorological tower situated at the S.C.K./C.E.N.

As well as air-temperature T, wind speed ff and direction dd are measured at different heights h along the tower. Table 1 summarizes the instrumentation of the tower.

Table 1

h (m)	T	ff	dd
8	x	-	-
24	x	x	x
49	x	x	-
69	-	x	x
78	x	x	-
114	x	x	-

As the standard deviations ($\sigma_y(x)$ and $\sigma_z(x)$ of the cross-wind and vertical distributions of materials suspended in a plume are functions of the stability of the air in the friction layer, a stability parameter had to be adopted. At the S.C.K./C.E.N. the following stability index is used:

$$S = \frac{\partial\theta/\partial z}{\bar{u}_{69}^2}$$

with $\partial\theta/\partial z$ = potential temperature gradient between 8 m and 114 m (hourly mean)

$\bar{u}_{69}$ = windspeed at a height of 69 m averaged over one hour.

From the value of S, six stability classes can be defined, while a seventh class includes all the situations with an hourly average windspeed larger than 11.5 m/s.

For each stability class a functional relation between $\sigma_y(x)$, $\sigma_z(x)$ and S has been determined experimentally.

The available *computer facilities* for the pilot net are:
A PDP-8/E. 12K-12 bit core memory and 32K disk memory for on-line data acquisition, validation and reduction. As peripheral equipment an ASR-33 teleprinter and tape puncher, a power failure detector, a programmable real time clock and a high-speed paper tape reader/punch.

The central IBM-360 computer, will be used for the analysis and statistical treatment of the data. In the beginning this computer will work off-line but later on experiments on-line will be carried out e.g. for the use of mathematical models in alarm situations.

The software for data acquisition, validation and reduction by means of the PDP-8/E computer had been developed and is tested now. The meteorological instruments as well as the pollution monitors are interrogated every minute. Within a few milliseconds three successive readings per instrument are made and averaged if the dispersion is within a given range. This result has to pass several validation tests, such as a parity check, a range control and a gradient test, before being accepted and stored. After 30 minutes, averages are with the accepted one minute values and the complete cycle restarts. For the instruments with a built-in averaging time a special procedure is worked out. As well as the sampling interval the averaging time can be changed and the individual measurements are available, during a limited period function of the available mini-computer memory, for further analysis on the IMB-360.Other tasks of the mini-computer are: the telecommand of the instruments, organisation of the measuring and calibration cycles and so on. More details of the many different functions of the PDP-8/E are given in [3].

As to the *software for the central computer* mathematical models have been developed [4] and will be tested and improved by means of practical measurements. Work is going on to develop the software for the analysis, interpretation and optimum use of the information from an automated network.

The test programme for the *monitors for gaseous pollutants* consists of two parts. In the first one several monitors for different pollutants (SO_2 and NO_x at the moment) are examined under laboratory conditions. The tests are carried out with a calibration unit, based on permeation tubes as sources, allowing the separate or simultaneous injection of one or more pollutants in variable concentrations at variable temperature, relative humidity and flow rate.

These parameters are changeable between the lowest and highest operational limits of the instruments. In this programme the operators will have the opportunity to become familiar with a variety of instruments of different measuring principle, conception and design. The instruments selected after the laboratory tests will go to the second set of tests where they have to work under the conditions encountered in an automated monitoring network. This means that in this phase they are under the supervision of the PDP-8/E while a further check of their availability, reliability and flexibility, under normal and extreme conditions, is made.

For *dust monitors* a similar programme is going on. A comparative study between fundamentally different instruments has been started. Same of the instruments under investigations are a Nephelometer, a Particle Mass Monitor and a β-absorption Mass Monitor. Each of these instruments has its specific range of concentration and averaging time. The forms of the output signals are quite different too and this causes special problems for the cimputer hard- and software.

On-line monitoring of radio-active air-pollution has not been started yet although the α, as well as the $\beta-\gamma$ activity, is checked continuously by means of six stations around the S.C.K./C.E.N. site [5]. Once some experience with the system for non-radioactive air-pollution has been acquired the existing network for radio-active pollution will be automated and included in the Pilot network.

CONCLUSIONS

The scientific problems, technological difficulties and costs, characteristic of the installation and operation of large automated air-pollution monitoring networks require a profound study of the concept as well as of the system proper before a final choice is made. This, together with the changing nature of air-pollution, the broadening of our scientific knowledge and the improvement in the technology of monitoring, were arguments enough for the Nuclear Research Center in Mol to construct a Pilot net in order to have a scale model and research unit where a wide range of problems could be simulated, studied and solved in a flexible and dynamic way.

The first steps towards the realization of the objectives have been made and the results will have to prove the soundness of our approach.

ACKNOWLEDGEMENT

Thanks are due to Mr.X.de Maere, project leader, and all the members of the research team Pilot net for many interesting discussions and suggestions.

REFERENCES

1. S.C.K./C.E.N. "Automatische Luchtverontreinigingsnetten - Antwerpse Havengebied en Gentse Kanaalzone - Voorbereidende studie". Maart 1972 (Restricted)

2 H.BULTYNCK, L.M.MALET, *Tellus XXIV (1972), 5.*

3 H.BULTYNCK, J.BONNIJNS, G.VAN ROOSBROECK, "Acquisition, Validation and Reduction of the data of automated monitoring stations", International Symposium on Air Quality, RIV, Bilthoven, 1973

4 G.COCQUYT, J.F.DE GREEF, J.VANDERVEE, "Data Management for the Computation of Urban Pollution Models", International Symposium on Air Quality, RIV, Bilthoven, 1973.

5 R.BOULENGER et al, "Contrôle Radiologique de l'Environment du C.E.N. `a Mol", Jan.1964, Rapport R 2298

ACQUISITION, VALIDATION AND REDUCTION OF THE DATA COMING FROM AUTOMATED MONITORING STATIONS

H.BULTYNCK, J.BONNIJNS AND G.VAN ROOSBROECK
S.C.K./C.E.N., Mol, Belgium

OBJECTIVES OF AN AUTOMATED AIR QUALITY NETWORK

Knowledge of the main objectives of an automated network is essential for the comprehensive setting up of the software in the Regional Data Reduction Center, which has to handle the Remote Terminal Units in a master/slave relationship. The main objectives of concern can be formulated as follows:

a. The data have to be gathered in physical units over fixed integration intervals (gathering function).
b. The operational performance of the measuring devices has to be checked regularly to be sure that the measured values are representative of the encountered concentrations of physical quantities in the atmosphere (calibration function).
c. The gathered data have to be permanently stored for further treatment (logging function).
d. A software treatment of the gathered data must direct the surpassing of preset values (alarm function).
e. All information about failures of the essential components of the system has to be displayed for easing maintenance (error function).

The operational structure of a network is built up in three steps:

- The Remote Terminal Unit (RTU), where one of up to six chemical components can be measured; some of them can deliver 16 analogue outputs (mammoth stations provided with complementary meteorological information);
- The Regional Data Reduction Centre (RDRC), provided with an on-line computer, which has to process the measured values into the definite form in which they will be stored for further interpretation;
- The National Data Processing Centre (NDPC), also provided with an on-line computer, where in a nearly continuous mode the values are stored under their definite form. This Centre has also to provide the necessary meteorological back-up in order to issue regular forecasts in normal pollution conditions or special forecasts under alarm conditions. This Centre can also be called upon to give evidence on the history of already-recorded

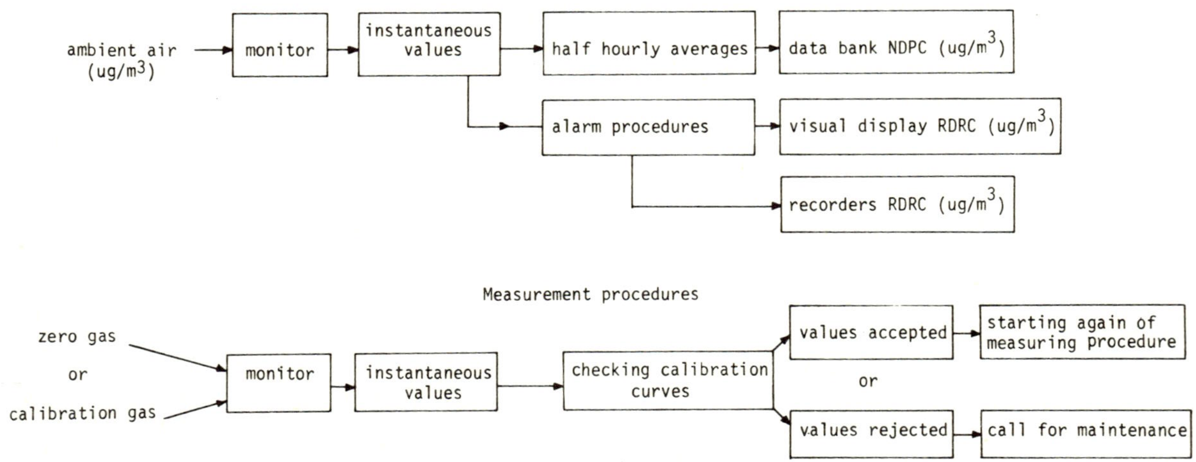

Fig. 1 - Calibration procedure

values. The software required for storing the data coming from different RDRC, for mathematical modelling with actually measured values and for the alarm forecasts and procedures is not within the scope of this paper.

ACQUISITION OF DATA

Two sets of data can be gathered: one set is related to actual measurements, the second one is connected with calibration values (Fig. 1)

Normal measuring procedure

The values representative of the encountered levels of pollutant concentration and of meteorological quantities have to be related to a choosen integration interval. One will adopt half hourly averages as a standard for air pollutants. To benefit from the highest possible repetability, each monitor will be interrogated with a sampling interval of one minute. When one has to build up half hourly averages with one-minute values, due account has to be taken of the physico-chemical time constant of the monitor and a filtering technique has to be applied either through analogue or digital means depending on the accuracy wanted for the average value, on the characteristics of the fluctuating signals to be measured and on the dynamic response of the measuring device. A higher sampling rate seems to be excluded for reasons of the size of the network looked after by an RDRC, of the speed of the modems (200 bauds) used for the teletransmission, of the amount of work to be performed by software at the RDRC during a limited time interval.

For the normal measuring procedure, provisions are made to convert the average values in electronic units into physical units by making use of characteristic equations relating output signals to the actual pollutant concentrations.

A special provision has to be made to activate the one-minute values to use them for feeding a visual display panel or analogue recorder , when asked for.

Calibration procedure

At certain intervals (e.g. once a day), the monitor will sample zero gas or calibration gas. The values will be activated by the operational software at the RDRC. The one-minute values will be stored temporarily to check the characteristic equations used for the translation into physical units and to check the physico-chemical time constant for each monitor. If the values

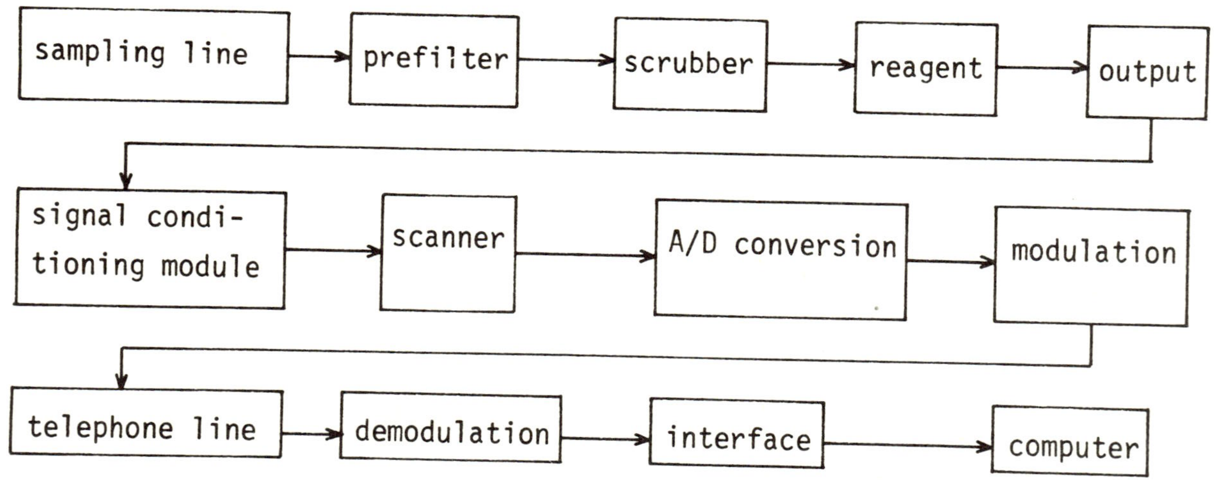

Fig. 2 - Diagram of the monitoring and data acquisition system

are located between preset limits, the measuring cycle will be activated, if they are outside the preset limits a call for maintenance will be issued in plain language.

The calibration procedure is of a vital importance to ensure the degree of reliability of the different parts of equipment used in a monitor: prefilter (permeability, ad- and desorption), scrubbers (efficiency), reagents (reaction time, degree of transformation). For the time being, the test is conducted in a yes- or no-fashion. If in the meantime the calibration devices prove themselves sufficiently reliable, a procedure of correcting the characteristic equation by a two- or three-point calibration could be envisaged.

VALIDATION OF DATA

One of the main features of an automated network is the enormous amount of data which is continuously built up. The philosophy for the use of the databank is of vital importance and is a matter of concern of the NDPC. As the data will be used for statistical treatment, for investigation of the air quality in different areas with respect to the exposures of the population, for general studies of modelling air pollution, for correlating concentrations with meteorological data or for intercorrelating measurements of different pollutants, it is also of primary importance to store data which are entirely reliable. Therefore one must call upon severe validation techniques for checking data before they are released to the data bank. As the raw data are only available at the RDRC, special software provisions have to be made to realize sophisticated validations. Checking the characteristic equations by calibration provisions (zero- and calibration gas) is already the application of a first hardware validation principle. Calibration techniques check most of the physico-chemical procedure (Fig. 2) except for the sampling line itself. Once the electronic signal is realised at the output, all kind of noise can intervene so that the signal message can be distored.

The steps in the proposed validation procedure can be listed as follows:

1. *Test on the transmission of the messages between the RDRC and the RTU*

   Three tests are carried out in the following logical sequence:

- Does the RTU react when interrogated by the RDRC?

  When the RDRC has not received an answer within a preset time, a new interrogation procedure is activated. After three times without response the interrogated channel within the RTU will be disactivated in the acquisition program with a special mention in plain language for a call on maintenance.

- A software parity check is done before accepting the message coming from the RTU.
- The address of the RTU and the position of the scanner are verified.

2. *Test for the acceptance of the incoming electronic signal*

For each channel the incoming signal must lie between two preset values. If they do not, they are rejected.

3. *Test for the acceptance of the individual measurement*

To be sure that the incoming electronic signal is representative of a real measurement each channel will be interrogated as quickly as possible several times (in practice three times) one after the other. The second and the third incoming values will be compared with the first one. If the difference is higher than a preset value, taking into account the fluctuation of the quantity to be measured, the dynamic response of the measuring device, the time constants imposed by the electronic circuitry, than all three values will be rejected.

4. *Tests for the acceptance of the individual measurement in relation to the measurements belonging to the same integration interval*

To see of an individual measurement belongs to a group, one could compare its deviation to the mean with the standard deviation of the group. This technique cannot be used in the present case because of the absence of any knowledge about the encountered distribution of the values. Therefore one uses a simple gradient test to see how far an individual measurement compares be succesfully adopted knowing the dynamic characteristics of the incoming signals, as it has been determined by the former test.

A special output has to be provided for these validation checks. This information will be of vital importance for the daily maintenance of the network and mostly will give some hints about failures of the system. A statistical treatment of the errors is also interesting for defining the downtime of the system and for taking corrective measures where needed.

It is the intention of the S.C.K./C.E.N. to develop the general software program of the RDRC in different steps. A first effort has been devoted to the data gathering of its meteorological tower, a second effort is being devoted to the data gathering of four RTU, one of them being an experimental monitoring station for atmospheric pollutants. The experience gained by these two efforts will lead to the editing of the operational software to be used in the final RDRC configuration.

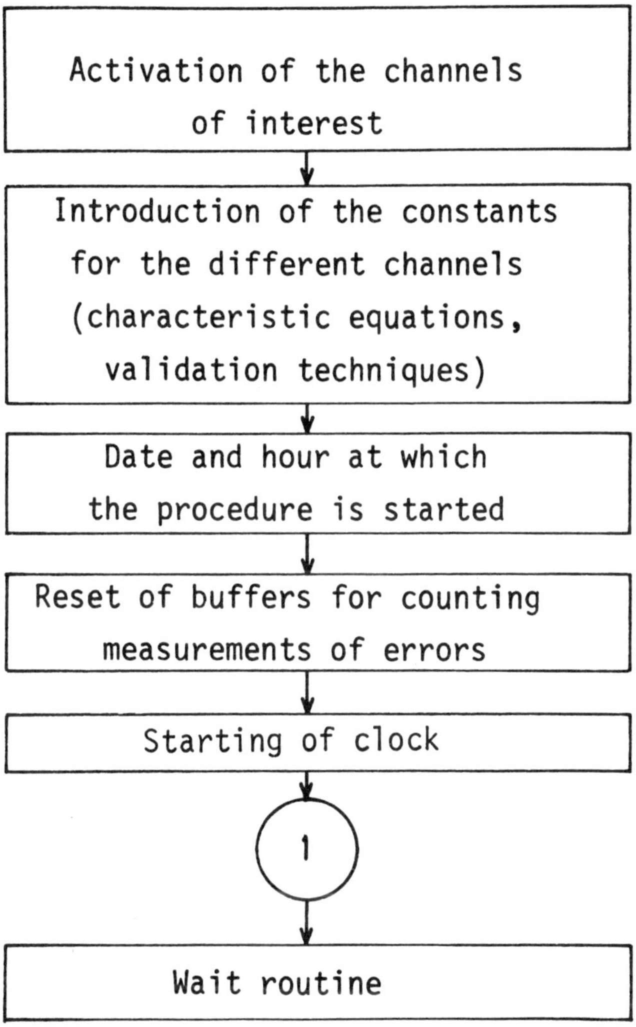

Fig. 3 - Start and initialisation

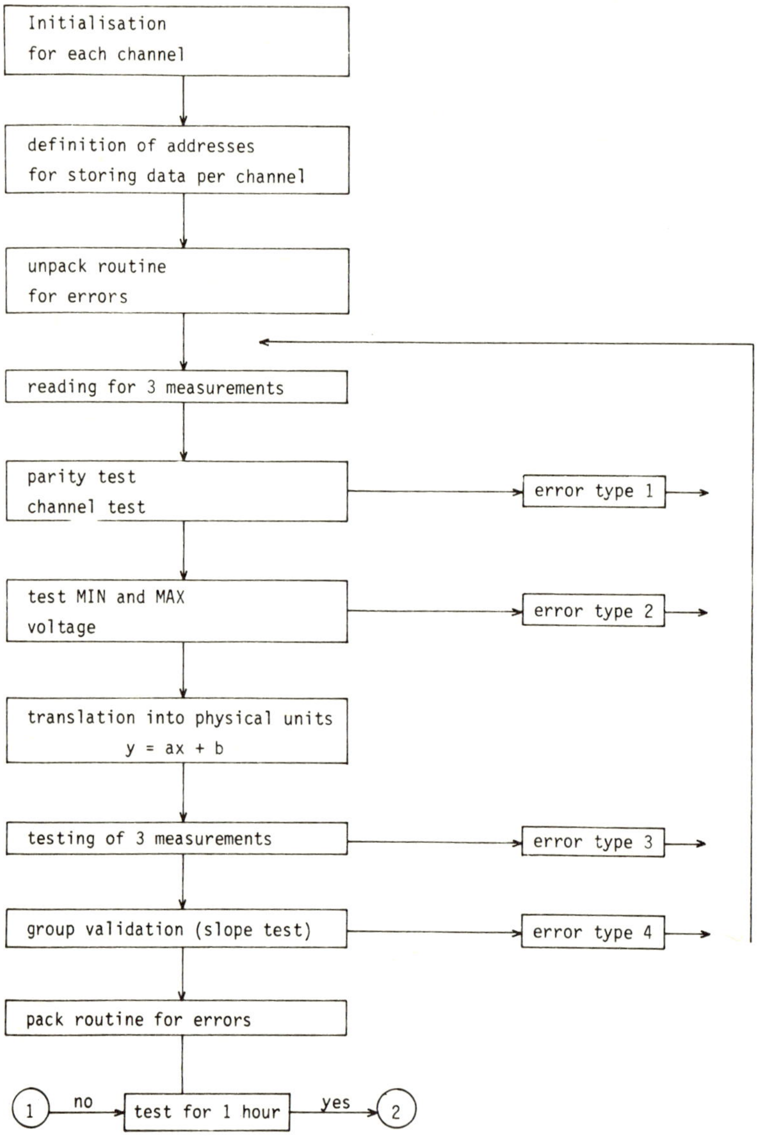

Fig. 4 - Data acquisition

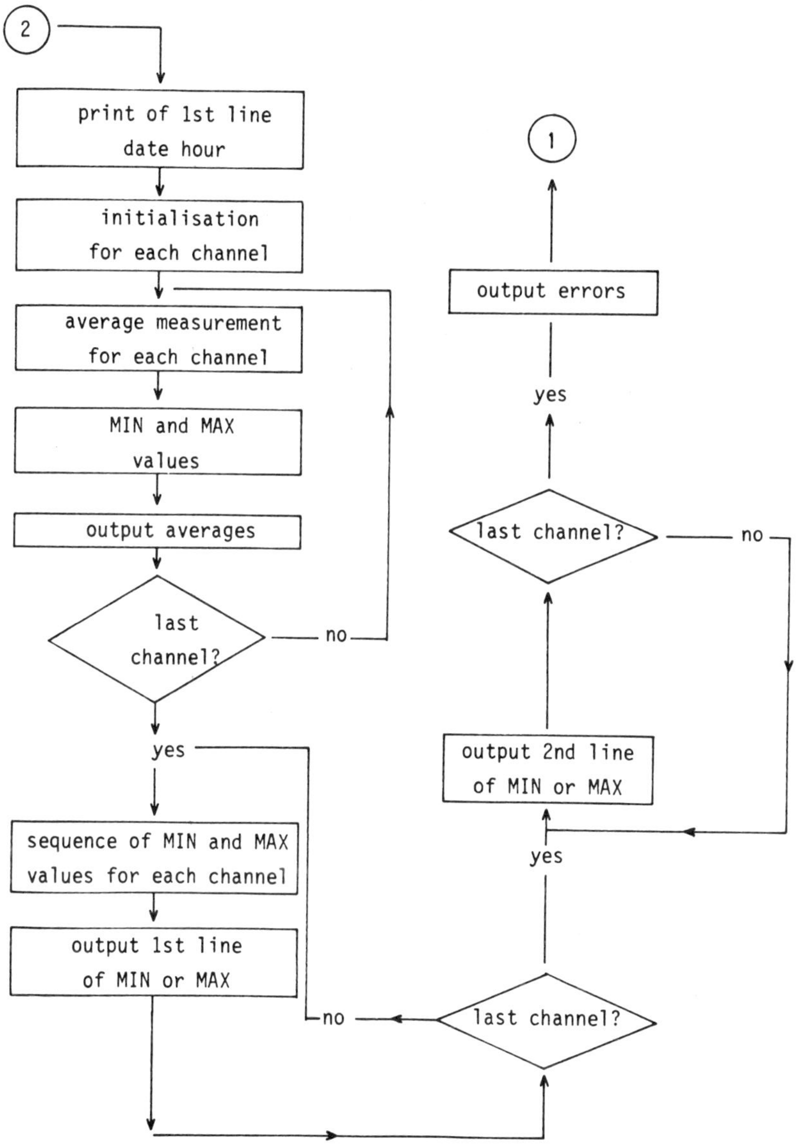
2
print of 1st line
date hour
initialisation
for each channel
average measurement
for each channel
MIN and MAX
values
output averages
last
channel?
no
yes
sequence of MIN and MAX
values for each channel
output 1st line
of MIN or MAX
no
last channel?
yes
output 2nd line
of MIN or MAX
last channel?
no
yes
output errors
1

Fig. 5 - Output

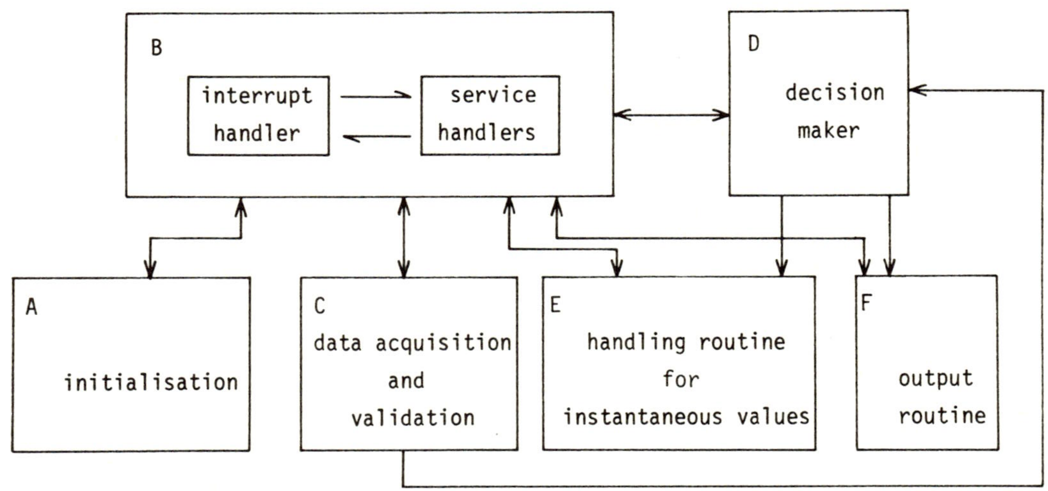

Fig. 6 - General organization of the software

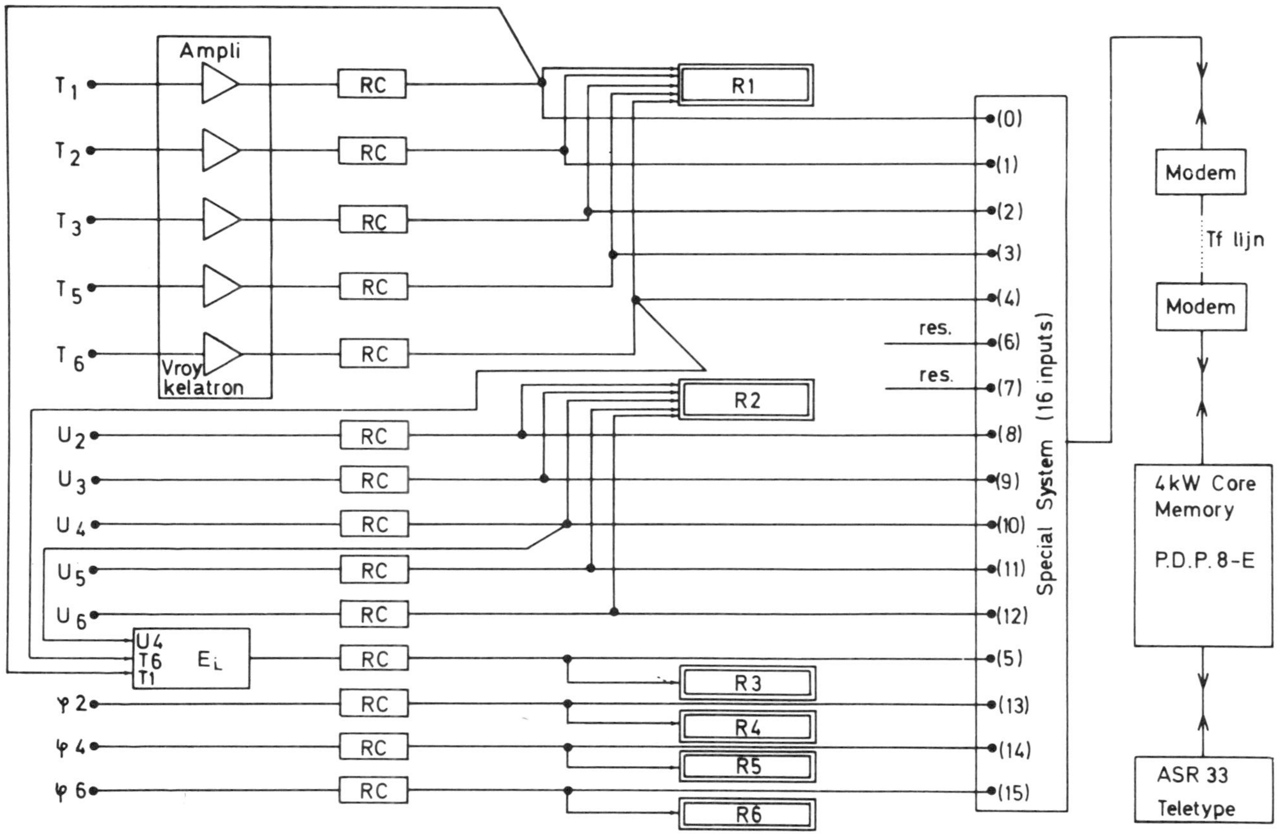

Fig. 7 - General hardware configuration

```
&0 03622100
+0352+0337+0344+0430+0605 013 034 078 080 119 116 150
+0279+0254+0260+0303+0545 010 054 063 072 088 108 142
+0406+0400+0418+0509+0648 020 020 089 093 131 128 154
 0000 0000 0000 0000 00000000000000000010001100000000&
```

T_1 T_2 T_3 T_5 T_6 E_i u_2 u_4 u_5 u_6 φ_2 φ_4

```
&0 03622200
+0495+0493+0507+0584+0657 015 034 071 081 108 133 155
+0418+0406+0420+0513+0610 010 041 057 062 099 120 150
+0582+0590+0612+0685+0709 020 024 088 095 117 141 156
 0000 0000 0000 0000 00000000000000000000000000000000&

&0 03622300
+0717+0706+0697+0701+0698 020 046 085 086 111 147 157
+0618+0624+0636+0672+0709 020 064 051 060 130 141 161
+0747+0733+0715+0715+0691 020 023 109 112 092 154 155
 0000 0000 0000 0000 00000000000200120013000100000000&

&0 03630000
+0758+0738+0717+0710+0697 024 051 082 082 104 146 156
+0745+0721+0709+0721+0691 020 037 104 097 126 156 158
+0763+0745+0725+0697+0703 030 063 063 064 078 138 154
 0000 0000 0000 0000 00000000000100000001000000000000&

&0 03630100
+0737+0718+0704+0704+0695 020 038 063 071 089 134 152
+0757+0739+0715+0715+0689 020 067 101 101 121 147 157
+0678+0672+0681+0691+0703 020 019 037 045 065 124 147
 0000 0000 0000 0000 00000000000100100002000200000000&
```

Fig. 8 - Illustration of output

EXAMPLE OF APPLICATION: DATA GATHERING FROM THE METEOROLOGICAL TOWER

The software modules for the acquisition, validation and reduction of data have been written on the smallest possible configuration of a mini PDP8-E computer: 4K-core memory and one ASR-33 as dialoging and ouput unit. The experience gained was interesting from the point of view of developing a program using the smallest possible core-size and applying the smallest possible execution time. These two characteristics will be again of vital importance in the software used in the large automated networks where extensive work has to be carried out with a relative small core- and disk memory in the shortest possible time interval.

The hardware configuration is given in Fig. 7.

T_K : temperature measured by means of thermocouples. Output signal: 0-10 volts DC

μ_K : wind speed measured with DC-generators housed within cup anemometers. Output signal: 0-10 volts DC

ϕ : wind direction measured with vanes delivering a DC signal of between 0 and 9 volts

K = 1 h = 8 m
K = 2 h = 24 m
K = 3 h = 48 m
K = 4 h = 69 m
K = 5 h = 79 m
K = 6 h = 114 m

E_i : atmospheric stability indicator, delivering discrete values corresponding to the stability class: 1 volt = E_1
2 volts= E_2
:
7 volts= E_7

The output comes from a hybrid calculator, developped by the Applied Mathematics of S.C.K./C.E.N.[1]. The input is delivered by T_1, T_6 and U_4. The device calculates the difference between two temperatures and the square of the wind speed. The relation between the two values is discriminated by comparators and a signal lying between 1 and 7 volts is generated as a function of the detected atmospheric stability class.

R_K : recorders for visual display

RC : elementary signal conditioning modules restricted to RC chains

Special system: This electronic device is provided with 16 analogue inputs (0-10 volts DC) and 4 digital ouputs, activated by the PDP8-E, in order to carry out 4 commands, independently and simultaneously.

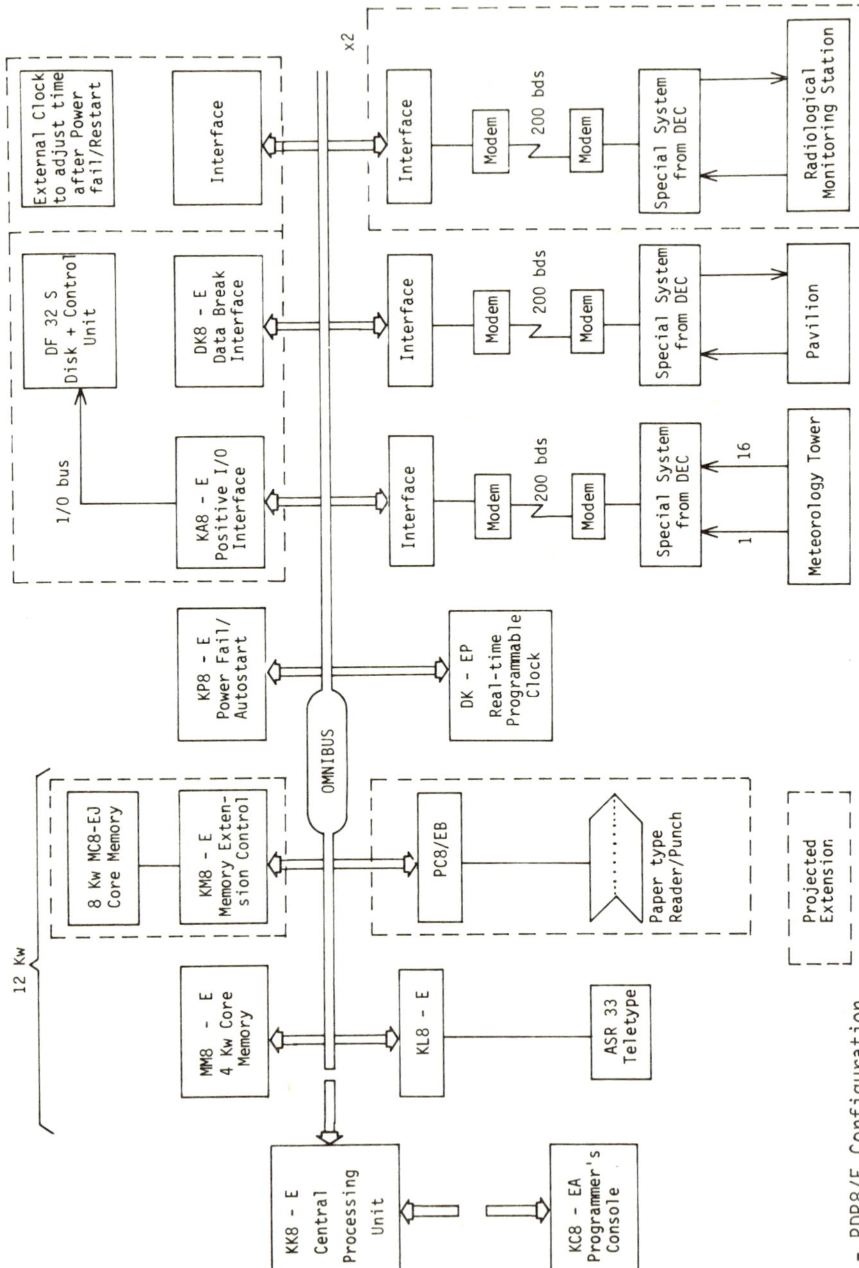

Fig. 9 - PDP8/E Configuration

The A/D conversion is carried out with a precision better than 0,1% F.S. This device has been tailored by DEC.

Modem: A 200 bauds KOOPMAN LEM/NR is used.

All devices are housed in a shelter at the foot of the meteorological tower and are situated at a distance of 150 meters from the PDP8-E configuration.

The general organisation of the software is given in Fig. 3, 4 and 5. The sampling frequency is fixed at once every five minutes, the integration interval is fixed at one hour.

An illustration of the output is given in Fig. 8.

The first line gives an indication about the identification

e.g. & 03622100

&: meteorological tower

0: no power failure. If ≥ 1, power failure has occurred and the dating block has to be corrected because of the down time of the system.

0362: xth day of the year: 362

2100: 2100 h (z).

The second line gives the average hourly values for T_1, T_2, T_3, T_4, T_5, T_6, E_i, U_2, U_4, U_5, U_6, ϕ_2, ϕ_4.

T_K reads in 0.01°C e.g. + 0352 = + 3.52°C

E_i reads in tenths and must rounded off in whole number e.g. 013 means $E_i = E_1$

U_K reads in 0.1 ms^{-1} e.g. 034 = 3.4 $m.s^{-1}$

ϕ_K reads in degrees e.g. 116 = 116 degrees

The third and fourth line give the maximum and minimum values in the order they appear.

+ 0279 } means that the maximum temperature in the past hour was + 4.06°C and
+ 0406 } the minimum temperature was + 2.79°C. The minimum temperature occurred first.

The last line gives the error count for 4 validation techniques:

1. parity check and answer from the right channel;
2. check of the output interval;
3. check of the individual measurement;
4. group test by slope calculation.

Zero means no errors. From 1 to 6 means 1 to 6 errors. 7 means more than 6 errors; in such a case the average value is not calculated, in the measurement area crosses appear.

EXAMPLE OF APPLICATION: INTERROGATION UP TO 4 RTU

Making use of the software modules, basically written for the acquisition and validation of data coming from the meteorological tower, a new program is being written for the treatment of data coming from 4 RTU, each provided with 16 analogue inputs, with an expanded PDP-8E configuration consisting of:

- 12 K core memory
- 32 K disk memory
- high speed paper tape reader and punch
- ASR 33 as dialog and output device.

The hardware configuration is given schematically in Fig.9.

The first RTU consist basically of the 16 analogue values issued by the meteorological tower, as described in an earlier paragraph.

The second RTU consists again of 16 analogue outputs from chemical monitors measuring atmospheric pollutants and/or by dust monitors. 16 or more digital inputs are foreseen for actuating remote by controllable parts of the monitors (e.g. valves). The sampling frequency is fixed at once per minute , the integration interval at half an hour.

The third and fourth RTU will consist of radiological monitoring stations. They will be operational in the course of 1974.

The general features of the program make it possible to preset the sampling frequency and the integration interval independently for each RTU. Two limitations are imposed by the system. The sampling frequency cannot be set higher than once a minute and the integration interval cannot cover more than 30 instantaneous values. Further, the initialisation of each handling program is independant of the RTU, so that the acquisition program for each RTU can either be disactivated or changed without interrupting the running acquisition program of the remaining RTU.

Two basically different programs can be run: the first one consists of data acquisition, validation and handling of averaged values over the imposed integration interval, the second one consists of data acquisition, validation and handling of instantaneous values. The second provision is provided with a special output of instantaneous values for checking calibration values. The second program can be called in, either by hardware or by software, at fixed intervals.

The general organization of the software program is outlined in Fig. 6.

1. *Block A: Initialisation*

   1. Activation or disactivation of a special system.

2. Choice of the handling program, either devoted to average values over a chosen integration interval, or devoted to the handling of instantaneous values.
3. Activation or disactivation of the output channels of a special system.
4. Bringing in the constants for the characteristic equations and validation techniques.
5. Indication of the number of elementary time intervals.
6. Definition of the duration of the elementary time interval.
7. Introduction of the starting time.

2. *Block B: General device handlers*

1. Interrupt handler: - power failure
   - clock
   - key board
2. Service handler : - clock
   - in- and output devices (special system, ASR33, disk memory, high speed punch and reader)
   - telecommands

3. *Block C: Data acquisition and validation*

1. Acquisition routine for each channel in each special system
2. Validation modules
3. Error treatment

4. *Block D: Decision maker*

Activation of parts of the program devoted to acquisition, or storage on disk, or handling, or issuing of telecommands or output.

5. *Block E: Handling routine for instantaneous values belonging to the same integration interval*

1. Group validation by slope test.
2. Averaging of the instantaneous values over the choosen integration interval.
3. Translation into physical units

6. *Block F: Output routine*

1. The instantaneous values are directed to the high speed punch.
2. The mean values are directed to the high speed punch and to the ASR 33.

Data stored on paper tape issued by the high speed punch are intended for further treatment by a large capacity computer.

REFERENCES

1 G.COCQUYT and A.HEUSE, *Indicateur de zone atmospherique*, Document interne S.C.K./C.E.N. 622-202/71-01 du 15 janvier 1971.

AUTOMATED SYSTEMS FOR AIR POLLUTION MONITORING IN THE NETHERLANDS

N.D.VAN EGMOND

Rijks Instituut voor de Volksgezondheid, Bilthoven, The Netherlands

OBJECTIVES

During the last decade the Netherlands have become fully aware of the problem of air pollution and its impact on the entire society, varying from inconveniences to serious danger for public health.

A few years ago a strategy was developed by which it should become possible to cope with this typical problem of the twentieth century.

The Netherlands are densely populated and heavily industrialized. Increase of energy consumption, traffic and industrial activity causes, notwithstanding a favourable climate, environmental problems on the whole and air pollution in particular.

Although air pollution is perhaps not the most serious environmental problem it certainly is one of the most difficult ones. As the phenomenon of air pollution extends in space and time, measurement strategy should meet the following qualifications:

1. Providing information about the existing concentrations at ground level. This requires information about the variations of the immission patterns, in time as well as in space. It should be known if certain standards, which are mainly based on frequency distributions of immission values, are exceeded. Long-term trend analysis should be made, although this is very complicated due to long-term meteorological influences. Nevertheless one has to establish the influence on these trends of the growth of industry, population and traffic, taking into account the effectiveness of countermeasures.Zoning decisions have to be made with respect to immission patterns and deduced trends. Measured concentration values at ground level are the basis of warning functions; higher degrees of pollution have to be detected in an early stage in which it is still possible to take countermeasures to control the emission. It will be clear that unfavourable meteorological circumstances play an important role in the prediction of higher pollutant levels.
2. Proving information about the quantity, composition and origin of the pollution, national as well as international. To achieve this, immission

data should be supplemented with information, obtained with mathematical models, emission inventories and/or additional emission measurements. Although the use of transport models is very attractive with regard to prediction of immission patterns, it seems to be more preferable to establish empirically the relation between emission, transport and immission of air pollutants.

THE DUTCH MEASUREMENT STRATEGY

It will be clear that for an environmental system of which one of the objectives concerns the detection of high pollution levels in an early stage, real-time and on-line processing of data is required. The described objectives are met in the following way:

A) Immission measurements should be made, using on-line, real-time systems and additional mobile monitoring units. In order to obtain information about the immission distributions due to remote sources it was found necessary to get data from the whole country using a National Centre for data collection and analysis. Moreover, it should be mentioned that a warning function for industrial and densely populated areas all over the country not only depends on the measured levels in the areas of concern but also on background burdens and the degree to which long term transport take place. After experiments with a small pilot net, there are now three automated monitoring networks in the Netherlands.

a. The Regional Network Rijnmond, which is the responsibility of the Rijnmond Authorities.

b. the National Monitoring Network for Air Pollution and

c. the Pilot Project "South-Holland".

The latter is a pilot net and the parameters of measurement of this network are nitrogen dioxide and nitrogen monoxide, particulate matter, carbon monoxide, ozone,individual hydrocarbons and individual organic and inorganic sulfur compounds. The system runs fully automatically and comprises six stations in the western part of the country.

The objectives of this automated system are different from those mentioned before; obtaining information about the composition of the polluted air, especially during episodes of a high air pollution, is given priority. Less attention is payed to the origin and the distribution of the pollution over the given area.

Finally mobile units complete the immission measurements; especially in episodes with a higher degree of air pollution they play an important role in the determination of local high levels, in combination with the National Monitoring Network. The use of mobile units meets one shortcoming of an automated network, namely lack of flexibility.

B) The structure of a monitoring network is two dimensional; this means that no information about the pollutant distribution in the third dimension (vertical direction) is obtained. It is however, possible to meet this problem partly by the use of meteorological towers. In Cabauw, located in the centre of the country a 200 m tower is fitted with three SO_2 monitors at levels 3, 100 and 200 m. The monitors are connected to the National Network and can provide information about transport over greater distances.

To establish the empirical relation between immission and emission, both in time and in space, one needs immission as well as emission data. The immission data are obtained by use of the automated network and emission data have to be gathered independently by inventories.

However, emission inventories obtained with inquiries have the disadvantage of some unreliability and time dependency.

Within one year many pollutant sources may disapear and others may by created.

Emission inventories based on stack measurements may underestimate emission quantities, due to the fact that both large and minor sources (leaks) contribute to the total emission quantity.

For these reasons, verifications of emission inventories are needed; measurements of total emission quantities, if they can be made, could be very useful in this respect.

Moreover, they would be very useful with respect to the control facilities at the disposal of the Executive, especially in periods with unfavourable meteorological circumstances.

That is why the Dutch Institute of Public Health started investigations into methods by which emission quantities can be determined using remote sensing methods. Laser techniques have been developed, but are still in the laboratory phase. Within a few months, however, operational use will be made of a Correlation Spectrometer (Barringer-Canada).

By integration of the molecular absorbtion in an optical path between the sensor and a remote source (diffuse skylight), the total pollutant burden can be measured as the numerical product of concentration and optical pathlength (ppm-m). From windvelocity, wind direction and total burden, measured in a vertical plane, the total amount of SO_2 and/or NO_2

crossing that vertical plane (kg/s) can be calculated. This created the possibility of making measurements of the quantities of sulfur and nitrogen dioxide emitted by remote sources. When used in an automobile, rapid measurements of emissions and transport of gas clouds can be made, and measurements like these prove to be an ideal combination with those obtained from an automated monitoring network.

STRUCTURE OF THE NATIONAL NETWORK

To meet the objectives described a monitoring network covering the whole country was designed. A small pilot net was used to get the required experience and know-how about the location of monitoring stations and to get an idea of the possible problems concerning the interpretation of data.

This pilot net (SO_2) was set up in the Eastern part of the country in Enschede, a city with 120,000 inhabitants.

Monitoring stations were located in and around the city. From the data obtained the following main conclusions were derived:

a. Monitoring stations in rural areas are not useful as an indicator of unfavourable meteorological circumstances. For these circumstances when atmospheric transport stagnates, the measured levels are fully dependent on local sources and if there is no local sulfur dioxide emission, no changes can be detected.

b. Rural stations are useful for monitoring over a longer range.The concentration due to the emission of the city was found to be 20% of its initial value at a distance of 25 km. Mean values for the city centre in the winter season were about 80 micrograms per cubic meter.

c. Monitoring stations in cities can be used as detectors for unfavourable sircumstances. Only for those cases in which a clear source-sensor configuration exits, changes in atmospheric stability can be measured in the centre of the city.

d. At lower concentrations the isopleths with the highest concentration values were displaced only one kilometer down-wind. So the city centre will show in almost all cases the highest pollution level.
SO_2 monitoring stations in cities should be placed in the very centre of it.

The Dutch National Network consists of two basic configurations;

a. A regular grid of monitoring stations all over the country. In N-S E-W direction the distance between the stations of the equidistant grid is

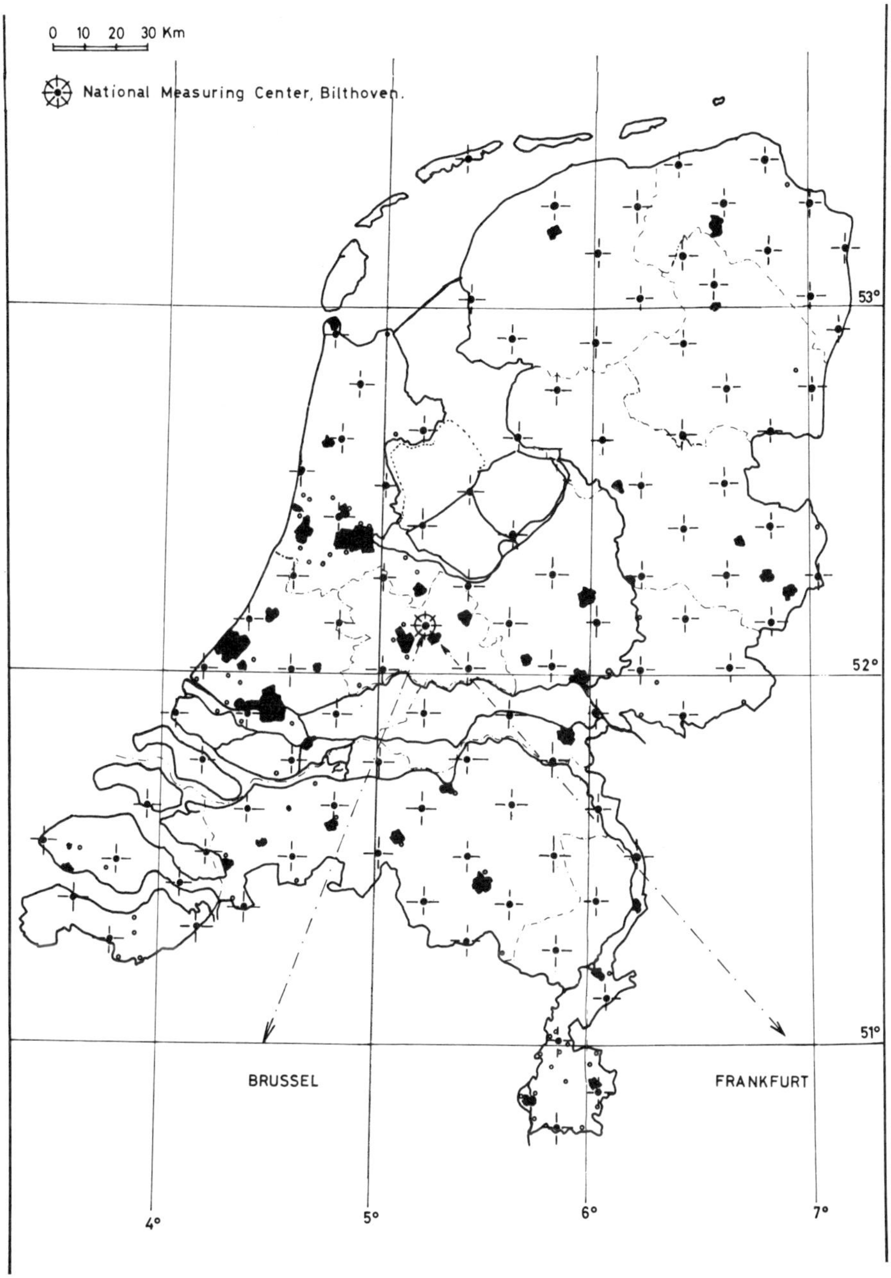

Fig. 1 - The National Monitoring Network For Air Pollution

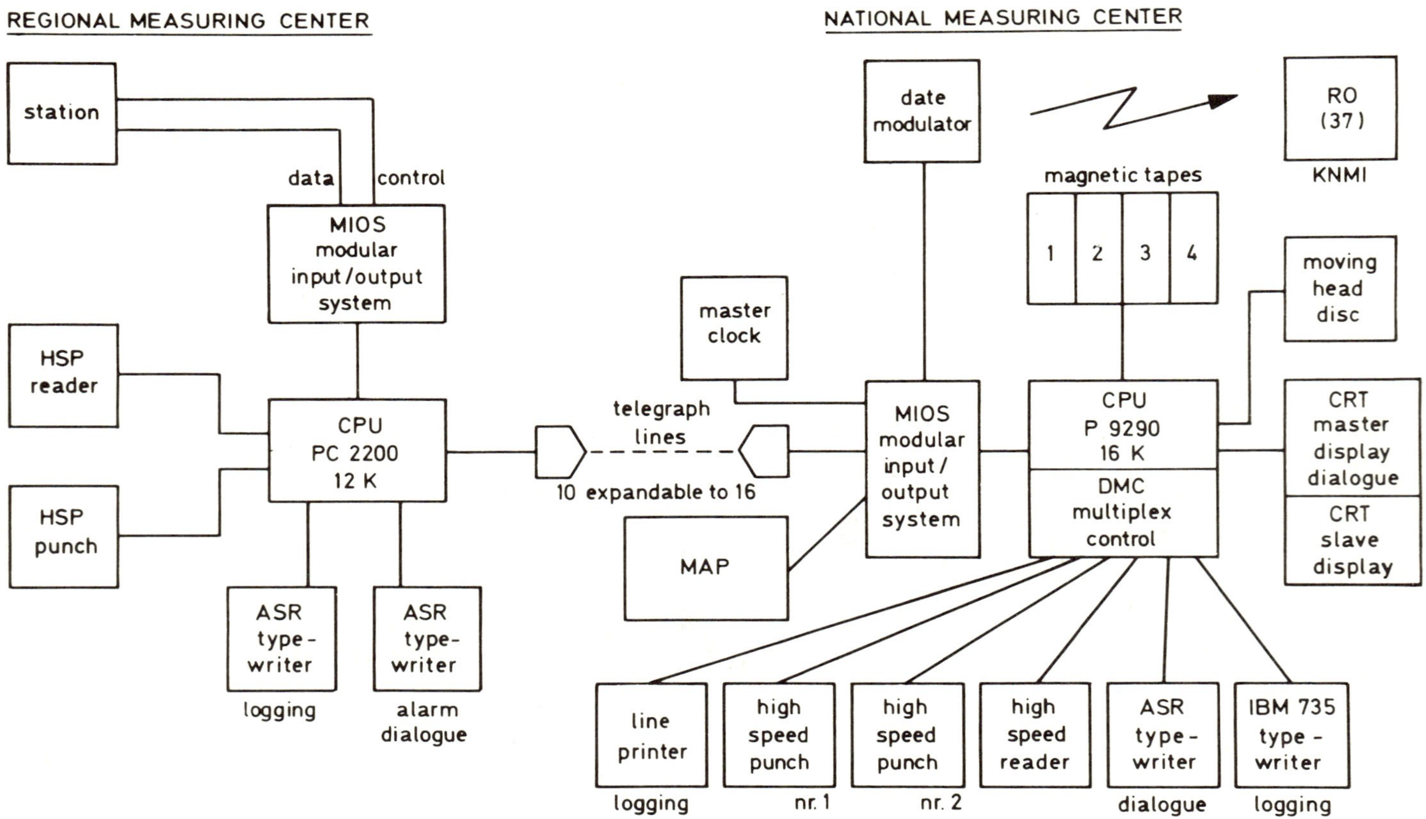

Fig. 2 - National Air Pollution System Configuration

28 km. Initial concentrations at citycentre of about 100 microgram are reduced to about 20 micrograms per cubic meter at a distance of 25 kilometers; this means that the influence of cities like this in general will be detectable. The Nordrhein-Westfalen Network in Germany is based on distance of 14 kilometers; to obtain a similar configuration the distance of 28 kilometers was chosen for the Dutch Network (Fig. 1).

The main importance of the regular grid concerns the required information about the immission distribution patterns in relation to the described objectivities.

b. Additional monitoring stations are located at borders and in densely populated and industrial areas, which are important in relation to the control and warning functions.

DESCRIPTION OF THE NATIONAL NETWORK

The National Measuring Network, now under construction in the Netherlands is composed of 3 essential levels: (Fig. 2)

a. Measuring Stations
b. Regional Centers, equipped with Regional Data Processors (R.D.R.)
c. National Measuring Center, at which the central coördination of the system takes place by the Central Communication Unit (C.C.U.)

The measuring stations are connected to the 10 Regional Data Processors by telephone line sending continuous analogue signals. After reduction and validation of data by the regional computer, digital information is send in blocks to the central communication unit in the National Centre.

MEASURING STATIONS

The SO_2 measuring station consists of a sampling arrangement, a chemical detector unit and an electronic unit for signal transmission to the Regional Data Processor. Samples are taken at a height of 3.80 m; after passing two filters in the sampler unit, the air is led to the measuring cell through a selective filter, by which components like H_2S, O_3 and Cl_2 are eliminated. The flowrate is kept constant by means of a pump-critical orifice configuration. The measuring cell contains a solution of KBr, H_2SO_4 and Br_2. The free bromine is reduced to bromide and SO_2 is oxidized to sulfuric acid: causing a redox potential change at an indicator electrode. A feedback system generates a current proportional to this change by which bromide is regenerated to bromine. This coulometric titration controls the bromine concentration at a constant level. The current required for the regeneration of bromine is recorded as the sulfur dioxide concentration in the air under observation, and as a signal in the scale of 0-20 mA converted to a frequency in the scale 5-25 Herz. This frequency is modulated on a carrier frequency by a tone transmitter and sent to the Regional Data Processor via a telephone line.

It should be noted that each monitoring station has its own carrier frequency and the Data Processor will recognize the station by this characteristic frequency. In normal operation the monitor will be in the measuring mode; zero check and calibration take place under control of the Data Processor. Every 12 hours a signal composed of two frequencies is sent to the measuring station and applied to two receivers via a line matching unit. This low frequency signal energize relays by which a motor controls the position of a tap. The positions of this tap correspond to the measuring, zero check and calibration modes of the apparatus. For zero check the tap is turned to a position in which the incoming air is led through a charcoal filter before entering the selective filter. When for the second time low frequency signals are received from the Data Processor the tap is turned another 90 degrees to the calibration position; a constant SO_2 source is switched between the charcoal filter and the selective filter.

The source consists of a glass reservoir filled with sulfur dioxide at atmospheric pressure and a permeation tube for delivering SO_2 at a constant rate.

From the zero-check and calibration signals received from a measuring station, the concentration for SO_2 in micrograms per cubic meter can be calculated by the Data Processor.

As described the measuring station contain a SO_2 monitor only but any required extension can be added without basic changes to the transmission system.

Meteorological parameters are processed in the same way as pollution parameters. Wind velocity and wind direction are processed; for wind direction two transmission channels are used for the sine and cosine of the angle of the vane to a given axis. In this way it will be possible to make an estimate of the variability of wind direction, a parameter which is expected to be very important for the interpretation of the imission data.

REGIONAL DATA PROCESSOR

About 20 measuring stations are under control of one Data Processor, a process computer used for primary data reduction. From the analoque signals received from the stations by telephone line, one minute values are calculated by counting the number of periods of these low frequency signals and correcting for zero check and calibration. From the one minute values, variable means, hourly values and daily values are calculated, and sent by telegraph line to the Central Unit.

The Data Processor can work in an autonomous mode or an automatic mode, selected by the National Measuring Centre. When operating in the automatic mode the RDR is under control of the Central Communication in the Centre and the local operator has only a restricted dialogue with the system. It is however, possible to print values from the measuring stations concerned or to track them on recorders. At twelve hour intervals the Data Processor supplies command signals for the zero check and calibration procedure of the monitors in the measuring stations. As described, these signals switch the tap in the monitor to the zero- and calibration positions. The signals obtained during these periods are stored in the RDR for reference during the next twelve hour interval. Software checks are made and possible failure of the calibration system of a monitor results in a message on the ASR dialoque typewriter. If the communication with the Central Unit is disturbed, the Regional processor switches back to the autonomous mode, by which the local operator is master of the system and the RDR timing system is used. In principle the regional system is designed to operate without the continuous control of a local operator; in case of a power failure an automatic restart procedure is available. However, most of the regional centres are manned and they can provide a direct local interpretation of the measurements, in collaboration with the National Centre.

THE NATIONAL MEASURING CENTRE

The Central Communication Unit (P9290-16k-16 bits) located at the National Measuring Centre has to provide control functions and to collect and dispatch data.

The CCU controls all regional data processors, if they are in the automatic mode; this means synchronizing the timing of the total system, zero check, calibration and processing, storage and presentation of information, obtained from the regional data processors (PC2200-12k-16 bits). On the NMC site two CRT displays and keyboards are available for operational control. Change of status (autonomous/automatic mode) of a region will be displayed automatically, important RDR failures will be printed on teletypes (ASR). Main information, namely the hourly values are logged on a line printer and stored on disc and magnetic tape for later, off-line processing. Hourly meteorological values are transmitted on line to the Royal Netherlands Meteorological Institute (KNMI).

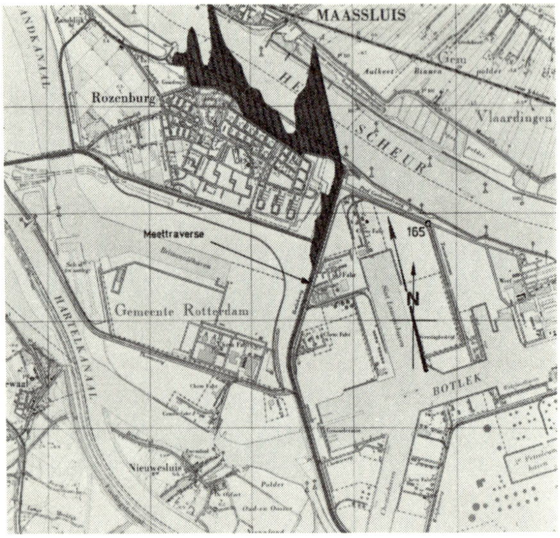

Fig. 3 - Total SO_2 burden in ppm.m measured by remote sensing mobile monitoring, in an industrial area

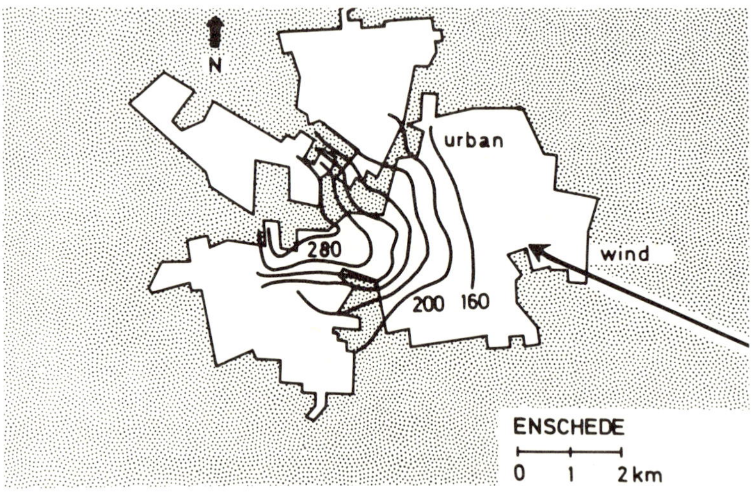

Fig. 4 - SO_2 immission distribution based on hourly values

They also are stored on magnetic tapes and displayed on the line printer and, if requested, on the CRT screen. Further on there are some functions operating on request of the Central operator; one minute and floating values from any station in any region can be displayed on several output devices like CRT and high speed punch (HSP). On recorders the specific values of a set of measuring stations can be tracked; possible errors in the measuring Centre has the control of the topographic display. The hourly values of all the available measuring stations are displayed on this map; lights of stations exceeding a chosen threshold value are switched on. By varying the threshold value it is possible to scan the pollution level in a few moments. This way of data display turns out to be very useful for subjective interpretation and control of the functioning of the measuring stations.

OFF-LINE DATA PRESENTATION

For the sake of clearness it has to be remarked that off-line data presentation should meet the initial objectives of the system. This means that the data have to be presented accordingly.

- frequency distributions, determined for every monitoring station; both distributions from data obtained with a given wind direction, and data obtained with different directions may be presented. If representative distributions of wind directions are taken into account, possibly in combination with other meteorological parameters, frequency distributions can be compared with given immission standards.
- Concentration-wind roses are useful for data interpretation; in combination with topographical data they can provide information about the origin and, in a measure, the quantity of the pollutant of concern. The accuracy of these estimates may be increased by using more detailed meteorological information.
- Immission distribution patterns (Fig. 4) based on absolute concentration values or 50% - and/or 90% values from frequency distributions, describe more or less the spatial distribution of the pollution.

  Although the interpolation of immission values between monitoring stations, without considering the source configuration in between, is not allowed, immission patterns obtained in this way inform the interpreter within a few moments about the ambient situation.

  Patterns are plotted by computer in a few minutes and parameters like wind direction and wind speed can be taken into account. The plotting of immission patterns based on hourly values can be seen as an extended use of the topographic display, mentioned before.

Data obtained by use of remote sensing techniques (Fig. 3) may be presented in combination with immission data from the National Monitoring Network. Till now remote sensing methods are not useful for measuring background transport rates and background concentrations; however high emission levels can be detected and in that capacity they complete the data obtained from the Network. Also in this respect the National Monitoring Network for Air Pollution will give a comprehensive insight into the Dutch situation in particular.

DISCUSSION

BURMANN

I would like some discussion on the terminology of calibration. In the U.S. to obtain one zero and span point would not be considered a calibration. A calibration would indicate a dynamic calibration with a primary standard throughout a complete range of the instrument.

BULTYNCK

The calibration that we are intending to do is a zero and a span calibration with gases passing through all the prefilters and through the aspiration parts of the monitor ,in order to see if the monitor is still responding well.Of course if possible, we can do more than a two point calibration, but we think that there is no problem with the linearity of the monitor. We think that the problem is zero drift, span drift and change of filter properties and this has to be checked.

BURMANN

Our experience has been that the only truly reliable standard that we can obtain is the SO_2 permeation tube. Even there we go back to the National Bureau of Standards in Washington to compare our commercial permeation tubes against the N.B.S. standards. However, it becomes very expensive, for example, to employ a permeation tube and maintaining this tube at a certain temperature and then get a microvolt calibration. Depending on the response of the instrument you may lose up to half an hour of your valuable data and even if you check the zero point on the instrument you are losing 10 minutes of data.

KRETZSCHMAR

I think one must consider calibration in the laboratory and calibration in the field. In a laboratory you can check the linearity of response. You can check it with a primary standard (permeation tube) and with a secondary standard because finally you will have to use that secondary standard in the field.

BULTYNCK

The secondary standard we will use in the remote terminal unit is only a yes or no indication. If the two point calibration shows that the original calibration curve is no longer applicable because we find values that are outside the accuracy range of the monitor, then the monitor is put out of use. In addition to the discussion on the loss of data during calibration I can say, since we derive half-hourly averages, at least one and a half hours are lost for each calibration. One half for zero, one half hour for span and then one half hour for the next measuring cycle, because of the physical-chemical time constant of the instrument.

KUMAZAWA

In Osaka we also have automatic span calibration, but we cannot rely on it completely. Since the stations are visited twice a week for maintainance we use these occasions to calibrate the instruments manually.

WHELPDALE

The detection of dangerous situations at an early stage seems to be one of the primary objectives of automatic real-time systems. Dr.Van Egmond mentioned dangerous meteorological situations often in this context. I agree that rates of change of pollutant concentrations are of interest here, but I think that meteorological forecasts should also be carefully considered for example pollution potential forecast. I think they would be helpful in giving much earlier warning than your atmospheric concentration measurements can do.

VAN EGMOND

Of course it is true that before a rise of the tracer gas concentration will be indicated, there must be a meteorological situation which influences this rise. As a matter of fact we know that anti-cyclonic air movements over Central Europe will give rise to circumstances in which the chance of getting an episode is very high. But I do not know what is the time span between the changing of the meteorological influences and the actual rise of the pollutant concentrations. We think that in an early stage, when the situation is expected to be very bad, the concentration levels detected in the area of concern will give information about what is going on there and still will describe the situation in the area. It is worthwhile to distinguish two forecasting stages: first a meteorological one and second the indication of the situation on the spot, and while we are alert because of meteorological indications, concentration levels are tracked to take action based on the actual rise of the real pollution concentration.

CLEARY

Can Mr.Nordø comment on forecasting and the accuracy that you might expect in air pollution forecasting generally?

NORDØ

It will be different from one place to another. For instance in Norway with its rugged topography, and in some other countries with mountains and valleys, it is rather easy to predict potential dangerous situations, because it depends much on cloudiness and wind strength whether you get an inversion or not. In central Europe the big episodes can be predicted well because it takes some time to establish a high. I do expect very large episodes to be predicted at least one, and very likely two days ahead with a certainty of 90%. The minor episodes depend very much upon the cloudiness distribution, due to the vertical velocity. At present, of the meteorological parameters the vertical velocity distribution is the most difficult one to predict in detail. So I would expect that minor episodes are more difficult to predict.

BERLIN

Yesterday Dr.Burmann indicated that he felt that for most of the automatic monitoring systems for SO_2 the level of the detection was about 60 or 70 $\mu g/m^3$. It was indicated that these automated systems can be used for alert purposes as well as for measuring the average values of the air pollution concentration. Since the summer values sometimes are lower than 50, 40 or even 30 $\mu g/m^3$, how can these data be used for such purposes, because how often do we find that the values obtained are below the detection limit. If one would go to this automated systems should one consider frequency distributions more than annual, winter of summer averages?

SHEARER

If you have single instruments or single sampling stations for collecting over infrequent or intermittent periods, simple frequency distributions and simple summary statistics are quite satisfactory. But if you do have continuous on-line monitoring stations, one should consider the use of more sophisticated summary and trend statistics.

DE KONING

I would like to come back to the question whether or not automated systems are required. In my opinion one has to make the distinction between off-line and on-line automated systems. I think that the choice between these should be based on cost aspects, avoidance of a lot of expensive labour, the necessity to have an alert system, etc.

SHEARER

In our continuous automatic CAPM-stations that we are running in the U.S., the data output from the analyzers goes both on to magnetic tapes and to stripchart recorders. The magnetic tapes are mailed to our facility in North Carolina where they go through the computer processing for the production of the summary statistics themselves. If during cases of air pollution episodes high levels are forecast a person is placed in that station to get read out of the information directed from the analyzers for assessing situations during the episode. We consider this an automatic type of station but not necessarily an on-line station.

RUFF

I think another argument for on-line systems might be the opportunity to know if the instruments are all working.

STEWART

I think this does raise the sort of questions that I was bringing up before as to the degree of sophistication into which we must enter in order to meet the minimum requirements. For the purpose that had just been mentioned, for example, we do not actually need a continuous on-line connection. We need on the one hand a facility of interrogation, so that we can in fact contact that CAMP-station by a telephone-call and on the other hand a short memory system which can be interrogated.

I would like to come back to the earlier speakers because they said that it was not terribly expensive to transmit data over lines continuously. We did find in a purpose study in the U.K. that it would certainly double the cost of a station to have it continuously connected in real time to a central computer; whereas the system of interrogation would be much cheaper to employ. We did have an excellent summary given us by Dr.Bultynck of the possibilities of the functions of monitoring stations. Some of these functions can, I think, be quite well fulfilled by local data acquisition. It may be we are in a stronger position in the U.K., that we have a fairly well organized local government system. We have organized the national survey that I described in my paper with the cooperation of these local authorities, so that we do have reasonably technically confident people and can have manual feed back. Maybe this is an unusual situation. I would like the discussion to broaden into some comment on the level of sophistication at which we have continuous, on-line,real-time, the level at which we have the possibility of remote interrogation, the level at which local data acquisition is adequate etc.

GREENFIELD

I would like to make a general comment about the questions raised by Dr.Stewart because they lead to the question that underlies the philosophy of how you plan to protect the environment. There is a basic difference between the approach in the U.S. and some countries in the Western World. We approach it from the standpoint of providing controls, in essence setting a standard and insisting that the standard be met locally by the application of controls. This is different from the type of approach which says: "you control intermittently by being aware of episodes" so we are not interested in on-line monitoring systems because that implies an intermittent type of control activity. We shall allow ourselves to determine over the longer period how well permanent control systems are really operated. When you decide to go to an intermittent control system then the question whether you go to an on-line system is a measure of how much you really expect emergency episodes to occur and how fine a control you have to exert.

BULTYNCK

A few comments on prices. The price of a RDRC amounts to 2,000,000 - 3,000,000 Belgian francs, without taking into account the rent of telephone lines. The price of a RTU is of the same order. With reference to the properties of systems, it is quite important that the system availability for each RTU is much higher in an on-line configuration than in a configuration with data stored on tapes, And that is because of the mechanical parts that are introduced.

GOOTJES

The costs of the Dutch monitoring system will be 3,000,000 Dutch guilders a year, all in. And as Dr.Van Egmond mentioned, that system consists of more than 200 on-line stations for sulphur dioxide. I expect when the complete network including NO, NO_2, O_3 and CO is installed, the annual running costs will be doubled, so they will be about 6,000,000 Dutch guilders a year. My expectation is that you can diminish the costs of the network by at most 1,000,000 Dutch guilders, but as a result of that reduction it will lose the important facit of being alert. In my opinion the costs of what is called a "sophisticated network" here are really low: every Dutch citizen will pay 70 Dutch cents a year, and then all is paid.

SHEARER

In the Belgian system one-minute values are used. That means half a million data points per year per analyzer. I want to ask Dr.Bultynck: are you really going to look at all those data points?

BULTYNCK

One-minute values are used, firstly to have a high repeatability in order to make half hourly averages in the RDRC, and secondly to validate the output of the stations.

ANALYSIS AND PRESENTATION OF AIR QUALITY DATA FROM THE WHO COLLABORATIVE AIR MONITORING NETWORKS*

GERALD G.AKLAND AND S.DAVID SHEARER

WHO International Reference Centre on Air Pollution Control, Environmental Protection Agency, Office of Research and Monitoring, National Environmental Research Center, Research Triangle Park, North Carolina, U.S.A.

AND

GRAHAM J.CLEARY

Division of Environmental Health, World Health Organization, Geneva, Switzerland

SUMMARY

This paper describes the data aspects of the WHO collaborative air pollution monitoring networks. Emphasis has been placed on the importance of the data analysis and evaluation, as this aspect is commonly overlooked in monitoring activities. In addition, the various interlocking relationships from sampling through analysis and data evaluation have been stressed. By properly recognizing the importance of valid data analysis, better quality information is produced that permits more meaningful trend assessment.

INTRODUCTION

The WHO program to assist Member Countries in establishing, developing, and operating air pollution monitoring networks was initiated toward the end of 1967. The program is implemented through the WHO reference centers and collaborating laboratories on air pollution, which include, at present, two International Reference Centers (IRC), the IRC for Clinical and Epidemiological Aspects of Air Pollution at the Medical Research Council's Air Pollution Unit, St.Bartholomew's Hospital Medical College, London; and the IRC on Air

---

*This document is a preliminary draft. It has not been formally released and should not at this stage be construed to represent policy. It is being circulated for comment on its technical accuracy and policy implications.

Pollution Control at the Environmental Protection Agency, Washington D.C., United States of America. In addition, there are three regional reference centers at Moscow, Nagpur, and Tokyo; seven national reference centers and eleven collaborating laboratories; plus laboratories in 19 cities in the Pan American Sampling Network.

One of the main functions of the reference centers and collaborating laboratories is to promote the use of uniform methods of measurement; to introduce reliable and effective procedures for the calibration of routine sampling and analytical method; to improve the quality of monitoring systems; and to standardize the handling, statistical analysis, and use of data. This program will also help in generating internationally comparable data on levels and trends of air pollution in some urban and industrial areas; these comparable data may help in identifying patterns of exposure to air pollution; may facilitate the planning and assessment of health effects studies carried out in different countries; and may facilitate the comparative evaluation of the effectiveness of national air pollution control programs — to mention only some of the uses of such information.

As a part of the program, a number of monographs on the measurement of common air pollutants have been prepared and distributed to the collaborating laboratories and national reference centers for use and comment.

WHO recently initiated a pilot study to evaluate and test a scheme for handling and statistically analyzing air pollution data. Forty-eight sites located within 16 countries are involved in this pilot study. The pollutants monitored in the study are confined to SO_2 and suspended particulate matter. Each country selected three sites within a major city, one from each of the three categories, "inner city - commercial", "inner city - industrial", and "suburban-residential", in order to be representative of existing concentrations throughout the city.

The *Data User's Guide*, written and distributed by WHO, explains in considerable detail the coding structure and method of entering required information on data forms to allow for computer processing of the data and subsequent preparation of various types of statistical analyses of the data.

This paper briefly summarizes the information contained in the *Data User's Guide* and discusses the general aspects of analysis and evaluation of air quality data.

DATA HANDLING AND ANALYSIS

A vital part of any monitoring system is the transformation of the collected, analyzed sample into a numbered quantification of the sample result. This procedure actually encompasses three sequential steps: (1) submitting the raw (unsummarized) data; (2) summarizing the validated data; and (3) reporting the results. The effectiveness of the data handling process is ultimately measured by the utility of the resulting data base for drawing correct conclusions about trends and for comparing stations within the monitoring networks.

In the following sections, the system for handling data is discussed. Each of the three aforementioned steps of the data system is presented with respect to the overall objective of generating internationally comparable data on levels and trends of air pollution.

RAW DATA SUBMISSION

Comparing concentration levels requires information about the *sampling location, method of sampling* for a specific pollutant (parameter), *method of analysis* for the specific pollutant, and the *units* in which the pollutant measurement is expressed.

Information about the sampling site is recorded on a *site form* as shown in Fig. 1. This information is required for each of the sampling sites and is usually submitted immediately after site selection has been made. If characteristics of the site change over time — for example, if the area surrounding the sampling site becomes more industrialized or populous — then a new form is submitted for up-dating the old with the more recent information. Only one site form is required even though more than one instrument may be located at the sampling site.

The first half of the site form is structured so that information can be recorded about the country and the political jurisdiction of the sampling site. Information such as city population, longitudinal and latitudinal coordinates, and time zone (difference from Greenwich Mean Time) is requested.

The second half of the site form is designed to provide more specific site information, which includes: the sampling site address; the sampler location with respect to center city, or suburbs; the dominating influence on the pollution levels (industrial, commercial, or mobile); and the elevation of the sampler above the ground and above mean sea level.

Fig 1: SITE IDENTIFICATION FORM

World Health Organization Site Identification Form

Mail To: Environmental Pollution Unit
Division of Environmental Health
World Health Organization
1211 Geneva 27, Switzerland

Form Completed By __________ Date __________ New ☐ Revised ☐

TO BE COMPLETED BY THE REPORTING AGENCY	DO NOT WRITE HERE
(A) ______ State ______ Population-oriented Surveillance Project	State / Area / Site: A ☐☐☐☐☐☐☐☐☐ (1 2 3 4 5 6 7 8 9 10)
(14-36) City Name (23 characters)	Agency: R (11); Project: 0 1 (12 13)
(37-51) County Name (15 characters)	
City Population (right justified) ☐☐☐☐☐☐☐☐ (52 53 54 55 56 57 58 59)	Time Zone ☐☐ (78 79); Action ☐ (80)
0 0 (60 61); Longitude Deg. Min. Sec. ☐☐☐☐☐☐☐☐ (62 63 64 65 66 67 68 69); Latitude Deg. Min. Sec. ☐☐☐☐☐☐☐ (70 71 72 73 74 75 76)	
Difference from Greenwich Mean Time ______	
(B) (18-78) Supporting Agency (61 characters)	State / Area / Site: B ☐☐☐☐☐☐☐☐☐ (1 2 3 4 5 6 7 8 9 10)
Supporting Agency, continued	Agency: R (11); Project: 0 1 (12 13); Action ☐ (80)
(C) (14-79) Optional: Comments that will help identify the sampling site (132 characters)	State / Area / Site: C ☐☐☐☐☐☐☐☐☐ (1 2 3 4 5 6 7 8 9 10); Agency: R (11); Project: 0 1 (12 13); Action ☐ (80)
(D) (14-79) ______	State / Area / Site: D ☐☐☐☐☐☐☐☐☐ (1 2 3 4 5 6 7 8 9 10); Agency: R (11); Project: 0 1 (12 13); Action ☐ (80)
(E) (14-38) Abbreviated Site Address (25 characters)	State / Area / Site: E ☐☐☐☐☐☐☐☐☐ (1 2 3 4 5 6 7 8 9 10); Agency: R (11); Project: 0 1 (12 13); Action ☐ (80)

(over)

Site Identification Form (continued)

TO BE COMPLETED BY THE REPORTING AGENCY	DO NOT WRITE HERE
(F)________________________________ (14 - 54) Sampling Site Address (41 characters)	State [F][ ][ ] (1 2 3) Area [ ][ ][ ][ ] (4 5 6 7) Site [ ][ ][ ] (8 9 10)
________________________________ Address, continued	Agency [R] (11) Project [0][1] (12 13)
Check the ONE major category that best describes the location of the sampling site. — Next, check the subcategory that best describes the dominating influence on the sampler within approximately a 1-mile radius of the sampling site.	Station Type [ ][ ] (55 56)
1. ☐ CENTER CITY — ☐ 1. Industrial ☐ 2. Residential ☐ 3. Commercial ☐ 4. Mobile	County Code [ ][ ][ ][ ] (57 58 59 60)
2. ☐ SUBURBAN — ☐ 1. Industrial ☐ 2. Residential ☐ 3. Commercial ☐ 4. Mobile	
	Elevation/Gr [ ][ ][ ] (72 73 74)
Specify units ________________________ Elevation of sampler above ground	Elevation/MSL [ ][ ][ ][ ] (75 76 77 78) Action [ ] (80)
Specify units ________________________ Elevation of sampler above mean sea level	

Data Forms

The *methods of sampling* and *analysis* for even the common pollutants, such as suspended particulate matter and sulfur dioxide, differ from country to country, and, for that matter, even within a country. The comparison of values between countries requires that common or comparable methods of sampling and analysis be used; for each data value recorded by a participating country, then, the method of sampling and analysis must be identified. A list of the more common methods of sampling and analysis and their accompanying codes are presented in Table I.

The *units of measurement* should be the same so that comparisons can readily be made. When the units of measurement differ from the standard reporting units chosen to present the data (usually $\mu g/m^3$), the data are stored in the original form but are converted to the standard reporting units for all summaries and reports.

The World Health Organization recognized the problems that might occur if attempts were made to collect data on all of the possible atmospheric pollutants that might appear throughout the world. They decided to focus attention initially on the previously mentioned pollutants, sulfur dioxide and suspended particulate matter, thus simplifying the data handling steps inasmuch as only two forms for submitting the data are necessary — the *hourly data form*, Fig. 2, and the *daily data form*, Fig. 3. The only difference between the two forms is that the hourly data form allows up to 24 hourly readings per day, whereas the daily form has space for only one reading, the daily average. The *hourly data form* is used for continuous automatic instruments, including the automatic SO_2 analyzer and the spot tape sampler. The *daily data form* is designed to handle 24-hour integrated data, such as total suspended particulate matter data, our 24-hour SO_2 data, such as data collected by a bubbler. Space is provided on the form for recording as many as four pollutants per day.

Prior to submitting the data to the International Data Bank, the data should be validated, a process that includes: (1) checking to see that the data characterize the ambient levels in an area or neighbourhood (representativeness); and (2) checking to see that the data are accurately coded and the calculations are properly done. Special attention should be taken to validate the extreme values observed during a month or year, because they, more than other values, are apt to influence, or bias, the summary statistics. Meteorological data (wind speed, direction, etc.) are extremely valuable for validating the high values, since high values occur most often when the sampler is downwind from a large source or when stagnation conditions occur (e.g., light winds accompanied with stable diffusion characteristics). Once they are coded and validated, the data are then submitted to the International Data Bank for statistical analysis and storage for future retrieval.

Data Accuracy and Precision

The significance of any data value is dependent on the accuracy and precision of the measurement. In general, accuracy and precision of the raw data are strongly dependent upon the sampling and analytical methodology employed and the capability of the person performing the sampling and analysis. The instrument must be in calibration and in proper operating condition before the sample results can be expected to be meaningful. Therefore, some routine maintenance and calibration, as well as an overall quality assurance program, are essential. In what follows, it will be assumed that the sampling and analytical instruments are calibrated and performing within the operating characteristics as specified by the manufacturer.

It is meaningless and confusing to attempt to obtain a degree of precision outside the limits of sensitivity of the analytical procedures. For example, if sulfur dioxide values are determined by a conductivity instrument and are reported to three decimal places, an investigator attempting to use these data can possibly make two erroneous assumptions: (1) the sensors are of a quality that such precise measurements are possible and (2) it makes a real difference whether the value is 0.391 or 0.394 ppm.

A similar problem can result when data read from strip charts or analytical balances are reported to an exaggerated accuracy. In evaluating such data, most people have a preference for numbers ending in zero or five. Consequently, in examining values read from a strip chart, one would find an inordinate number of values of 0.010 and 0.015 as opposed to values of 0.011, 0.012, 0.013 and 0.014. This can become a real problem if someone attempts to compare two numbers that could differ from each other primarily because of measurement procedures. To avoid this problem, a reasonable accuracy is specified for such readings. Individual readings should be made to one decimal place beyond this accuracy, and the results should be rounded off, using a consistent method, for reporting.

For the pollutants of present concern to the WHO, total suspended particulates and SO_2 concentrations should be reported in $\mu g/m^3$ without any decimals; e.g., 162 $\mu g/m^3$. (If sulfur dioxide is reported in ppm, then two places to the right of the decimal – for example, 0,01 ppm – is customary.) Spot tape sampling data (soiling index) should be presented only to the

FIG 2: HOURLY DATA FORM

WORLD HEALTH ORGANIZATION HOURLY DATA FORM

MAIL TO: Environmental Pollution Unit
Division of Environmental Health
World Health Organization
1211 Geneva 27, Switzerland

Less than 24-hour sampling interval

[1] ____________________ Agency

____________________ City Name

____________________ Site Address

Population Oriented Surveillance ____________________ Project

____________________ Parameter observed

____________________ Time interval of obs.

____________________ Method

____________________ Units of obs.

State (2–3) Area (4–6) Site (7–10)

Agency (11): R | Project (12–13): 0 1 | Time (14) | Year (15–16) | Month (17–18)

Parameter code (23–27) | Method (28–29) | Units (30–31) | DP (32)

Day		St	Hr	Rdg 1				Rdg 2				Rdg 3				Rdg 4				Rdg 5				Rdg 6				Rdg 7				Rdg 8				Rdg 9				Rdg 10				Rdg 11				Rdg 12			
19	20	21	22	33	34	35	36	37	38	39	40	41	42	43	44	45	46	47	48	49	50	51	52	53	54	55	56	57	58	59	60	61	62	63	64	65	66	67	68	69	70	71	72	73	74	75	76	77	78	79	80

FIG 3: DAILY DATA FORM

WORLD HEALTH ORGANIZATION DAILY DATA FORM

MAIL TO: Environmental Pollution Unit
Division of Environmental Health
World Health Organization
1211 Geneva 27, Switzerland

24-hour or greater sampling interval

[2] 1

Agency

City Name

Site Address

Population Oriented Surveillance — Project

24-Hour — Time Interval

State		Area			Site			
2	3	4	5	6	7	8	9	10

Agency	Project		Time	Year		Month	
R	0	1					
11	12	13	14	15	16	17	18

	Column 1	Column 2	Column 3	Column 4
UNITS OF MEASURE				
NAME	Total Susp.Part.	Sulfur Dioxide		
PARAMETER CODE	1 1 1 0 1 (23 24 25 26 27)	4 2 4 0 1 (37 38 39 40 41)	(51 52 53 54 55)	(65 66 67 68 69)
Method	9 1 (28 29)	(42 43)	(47 56)	(70 71)
Units	(30 31)	(44 45)	(58 59)	(72 73)
DP	(32)	(46)	(60)	(74)
Data columns	33 34 35 36	47 48 49 50	61 62 63 64	75 76 77 78
DP →	4 3 2 1 0	4 3 2 1 0	4 3 2 1 0	4 3 2 1 0

Day (19 20)	St (21)	Hr (22)				
0 1						
0 2						
0 3						
0 4						
0 5						
0 6						
0 7						
0 8						
0 9						
1 0						
1 1						
1 2						
1 3						
1 4						
1 5						
1 6						
1 7						
1 8						
1 9						
2 0						
2 1						
2 2						
2 3						
2 4						
2 5						
2 6						
2 7						
2 8						
2 9						
3 0						
3 1						

Y06000002 R01

SITE 002

YR/MO/DAY	suspended part. 111019101	sulfur dioxide 424019101
72/12/11	190	131
72/12/12	366	67
72/12/13	248	75
72/12/14	302	79
72/12/15		30
72/12/16		30
72/12/17		30
72/12/18	189	81
72/12/19	226	98
72/12/20	170	60
72/12/21	151	32
72/12/22		79
72/12/23		79
72/12/24		79
72/12/25	245	120
72/12/26	249	86
72/12/27	196	64
72/12/28	132	46

Fig. 4 - 24-hour data listing

tenths place, e.g., 2.1 COH/1000 linear feet.

SUMMARIZATION OF DATA

Before the data are entered or stored in the International Data Bank, an evaluation is made to determine if the data were collected from a site for which the prerequisite site information is in the bank. If not, the data are initially rejected. Rejections also occur when the accompanying information about the method of sampling, method of analysis, or the units of measurement are not included. These data are returned to the laboratory in question with a request that the missing information be provided.

After the data are entered for computer storage, they can be recalled in the raw data format, Fig. 4 (24-hours data) and 5 (1-hour data), or in summary format, Fig. 6. The raw data format is extremely useful for purposes of identifying atypical values. As mentioned earlier, these atypical values may be erroneous or they may reflect unusual meteorological patterns or source influence.

The raw data must be ultimately condensed into meaningful summary statistics because the volume of data even from one instrument becomes unmanageable. For example, over 8000 hourly values are produced by one automatic-continuous instrument each year. The statistics form the basis upon which judgments and decisions are made. Hence, it is important that the statistics reflect the "real" situation and not be greatly affected by biases. For example, a yearly average for particulate matter calculated from two daily observations would in no way represent the true yearly average for the site, just as a daily average of carbon monoxide calculated from values collected only during the morning rush hour would hardly reflect the true daily averages. The data must be relatively complete over the time interval of interest (for example, day, season, or year) so that such biases can be avoided.

Completeness of Data

Ideally, the raw data entering the computer should be checked for completeness. The criterion for completeness of continuous measurements should require that at least 75 percent of the total possible number of observations be present. During the pilot study, the data will not be screened for completeness, at least initially, in order to focus attention on the concentration levels occurring within a city. Later, however, in order to document trends validly, some criterion for completeness will be required

1-hour data listing

sulfur dioxide
4240113
n-v.1972

day	am 12	1	2	3	4	5	6	7	8	9	10	11	pm 12	1	2	3	4	5	6	7	8	9	10	11	daily mean	no.
1	40	40	50	40	34	25	25	35	35	40	65	145	185	225	185	165	160	160	180	145	85	150	160	165	106	24
2	160	160	190	190	160	95	100	120	260	75	50	45	40	30	30	30	35	40	40	40	40	80	40	30	85	24
3	35	40	45	60	70	90	65	110	105	75	65	55	40	55	60	50	60	60	70	70	80	75	75	60	65	24
4	55	35	35	35	33	25	20	40	240	250	175	170	235	235	265	335	215	140	105	140	55	40	35	30	123	24
5	30	25	30	35	25	30	50	70	150	135	85	80	110	95	40	40	40	45	40	40	30	25	20	20	54	24
6	25	25	20	20	20	25	30	40	45	40	35	35	30	25	20	25	30	25	25	35	25	40	55	45	31	24
7	25	25	60	75	70	90	100	90	145	140	115	130	175	135	85	75	65	50	45	40	40	30	30	20	77	24
8	20	20	20	20	30	35	155	185	250	150	65	65	80	145	150	225	155	105	105	75	60	65	120	100	100	24
9	90	65	50	90	105	75	65	95	395	485	285	260	295	275	285	240	140	130	125	185	135	85	70	70	159	24
10	80	75	55	65	75	135	125	165	230	240	175	125	105	120	145	115	105	115	90	85	75	80	55	60	110	24
11	60	45	55	50	55	90	60	65	90	100	115	100	85	90	75	50	40	80	30	25	30	30	35	40	60	24
12	30	30	30	30	30	25	20	20	35	45	35	35	20	25	35	35	30	35	30	35	35	35	35	30	31	24
13	30	55	90	115	95	85	125	150	130	135	125	120	150	140	165	160	35	45	65	50	45	45	35	20	92	24
14	30	20	25	30	20	20	55	80	75	55	90	75	45	65	95	110	115	85	70	80	90	105	100	85	68	24
15	70	65	65	55	35	30	20	20	25	85	45	50	50	40	40	45	55	65	55	40	65	55	45	45	46	24
16	60	65	75	60	70	80	85	90	100	55	85	70	75	80	95	100	125	85	75	65	65	50	60	85	77	24
17	90	105	130	120	135	110	135	185	185	80	55	60	70	95	165	180	140	155	155	140	120	140	70	45	118	24
18	30	25	10	20	35	85	10	110	275	380	40	10	10	25	30	10	10	10	10	10	10	10	10	10	49	24
19	10	10	10	15	10	10	15	25	35	55	25	25	30	25	70	75	175	125	200	195	110	95	90	85	63	24
20	30	80	75	70	70	85	105	125	150	135	125	135	205	155	180	180	155	185	115	120	100	105	110	75	120	24
21																										0
22											70	65	50	45	50	45	40	40	30	25	20	20	25	45		14
23	35	35	35	45	55	80	75	60	65	55	45	45	50	40	50	25	60	45	30	30	25	20	30	30	44	24
24	20	35	35	50	60	70	55	40	50	45	40	40	30	30	30	30	30	30	40	45	40	35	30	30	39	24
25	30	40	40	60	80	90	95	95	65	40	25	20	20	20	20	20	20	35	40	40	40	40	45	55	45	24
26	50	30	35	40	35	30	35	5	40	40	30	30	30	35	30	30	30	35	40	40	40	50	40	50	37	24
27	60	50	55	60	65	115	110	100	135	100	75	90	70	55	60	45	40	55	65	65	60	35	40	45	69	24
28	55	65	80	90	65	70	65	75	80	55	50	55	65	75	90	90	85	135	205	125	120	185	195	260	101	24
29	270	275	265	205	190	180	145	220	435	310	80	55	65	75	80	75	85	85	95	90	85	65	70	85	149	24
30	60	75	65	140	145	150	220	180	320	420	460	480	375	455	365	260	135	115	80	75	85	95	95	90	206	24
avg.	58	58	62	67	67	72	77	92	148	135	94	92	96	100	103	97	83	77	78	72	•62	63	63	62	83	
no.	28	28	28	28	28	28	28	28	28	28	29	29	29	29	29	29	29	29	29	29	29	29	29	29		686
max	270	275	265	205	190	180	220	220	435	485	460	480	375	455	365	335	215	160	205	195	135	185	195	260	495	

Fig. 5 - Concentration in ug/cu meter (25°C)
Instrumental conductometric

sampling interval: 24 hour　　units: ug/cu meter ($25^{o}C$)　　method code: 91

location	yr.	no. samp.	min.	10	20	30	40	50	60	70	80	90	max.	arith. mean	arith. std. dev.	geo mean	geo std. dev.
				frequency distribution, %													
ca8000001 r01	72	7	36	insufficient data									137				
ca8000002 r01	72	7	41	insufficient data									175				
ca8000003 r01	72	3	33	insufficient data									90				
cz4000002 r01	72	50	62	86	104	110	119	132	140	151	162	232	419	148	67.95	137	1.48
in1000001 r01	72	4	308	insufficient data									550				
in1000002 r01	72	4	279	insufficient data									542				
in1000003 r01	72	4	245	insufficient data									775				
it6000001 r01	72	58	12	16	22	30	40	45	51	60	68	83	110	47	23.12	40	1.77
it6000002 r01	72	60	7	15	18	26	33	40	55	60	65	83	116	47	26.98	38	1.96
it6000003 r01	72	28	20	40	57	63	63	70	77	84	106	114	149	77	29.09	71	1.52
sw4000002 r01	72	37	8	12	13	14	16	18	22	23	26	40	83	22	13.51	19	1.60
sw5000003 r01	72	55	10	13	15	18	18	20	23	28	31	43	71	25	13.68	23	1.58
y06000001 r01	72	31	86	138	170	189	205	207	233	257	412	422	490	254	*****	233	1.52
y06000002 r01	72	32	48	122	151	189	196	202	245	256	273	302	413	217	81.87	199	1.59
y06000003 r01	72	34	41	112	165	187	195	202	215	278	320	381	537	236	*****	211	1.67

Fig. 6 - Pollutant 11101 - particulate

Table I - Common Parameters and Methods

PARAMETERS	PARAMETER CODE	METHOD OF COLLECTION AND ANALYSIS		METHOD CODE
		COLLECTION	ANALYSIS	
Total Suspended Particulate	11101	Hi-Vol	Gravimetric	91
Soiling Index (COH/1000 LF)	11201	Tape Sampler	Transmittance	81
Soiling Index (RUD/10,000 LF)	11202	Tape Sampler	Reflectance	91
Sulfur Dioxide	42401	Instrumental	Pararosaniline colorimetric	11
Sulfur Dioxide	42401	Instrumental	Conductometric	13
Sulfur Dioxide	42401	Instrumental	Coulometric	14
Sulfur Dioxide	42401	Gas Bubbler	Pararosaniline	91
Sulfur Dioxide	42401	Gas Bubbler	Conductometric	93

Table II - Continuous Measurement Summary Criteria

Time Interval	Minimum Number of Observations
3-hour running average	3 consecutive hourly observations
8-hour running average	6 hourly observations
Monthly	21 daily averages
Quarterly	3 consecutive monthly averages
Yearly	9 monthly averages with at least two monthly averages per quarter

Table III - Urban Minimum Detectable Limits by Pollutant and Analysis

Pollutant	Analysis	Minimum Detectable Limit ug/m^3	Substitution Factor ug/m^3
Carbon Monoxide	Nondispersive infrared	573	286
Sulfur dioxide	Pararosaniline, bubbler	5	2.5
Sulfur dioxide	Pararosaniline, continuous	26	13
Nitric oxide	Colormetric	12	6
Nitrogen dioxide	Jacobs-Hochheiser	5	2.5
Ozone	Chemiluminesence	20	10
Total oxident	Colorimetric	20	10

Table II contains suggested possible requirements for the number of observations needed before summary statistics can be calculated.

For the data to be representative, it should be noted that a very important consideration is that the samples be collected throughout the time interval under study rather than bunched all together. Although the above check does attempt to draw attention to this, it does not reject or weight data if more samples are collected in one quarter of a year than in another. To obtain a meaningful, unbiased summary of data that have been collected unevenly over the year (especially data exhibiting seasonal variation), one would have to weight the data evenly according to the number of samples collected per season. For example, if 20 SO_2 samples were collected during the heating season and only 10 samples were collected during the non-heating season, the summary statistics should weight these seasons evenly for computation of the annual average. Such refinements to summary statistics are hardly necessary if the data are collected systematically or continuously throughout the year.

Values Below Detectable Limits

The problem of determining how to report values below the limit of detection of the instrument employed (in order that summary statistics can be calculated) is complicated by the fact that zero, the most likely value to be supplied, cannot be used when geometric parameters are to be calculated. This problem is handled by inserting a value equal to one-half the minimum detectable limit for each method and analysis technique. Use of the midpoint between zero and the detectable limit as the substitute value for concentration levels below the detection threshold is a reasonable substitute for zero, based on our experimences. To maintain consistency from year to year, one minimum detectable value is used for each pollutant even if the minimum detectable limit is changed, unless there is a change of one order of magnitude: for example, the detectable limit changes from 10 $\mu g/m^3$ to 1 $\mu g/m^3$. The current minimum detectable limits for several common pollutants are presented in Table III. The stubstitution value is the value used in summary reports when the computer encounters a value reported as below the detectable limit.

An additional point should be mentioned concerning the use of substituted values for values below the threshold of the method. When more than 25 percent of the measured levels are below the minimum detectable quantity, no statistics, such as means and standard deviations, are computed from the data. This constraint guards against the possibility of biasing the computed statistics. Furthermore, at least 50 percent of the measurements in a set of data must be above the minimum detectable concentration before a summary frequency distribution of the values is prepared.

PRESENTATION OF SUMMARIZED WHO DATA

Frequency Distribution

Frequency distributions constitute the primary method of data summary for the data. As can be noted in Fig. 6, in addition to the percentiles, information given in the summary includes: (1) number of observations, (2) minimum and maximum observations, and (3) arithmetic and geometric averages and standard deviations. The purpose of the information is to describe the reported data in a simple, straight forward manner.

Frequency distributions are prepared by first ranking the values in ascending order. Based on the total number of values, the percentiles are calculated such that the percentage of actual sample values is less than or equal to the concentration listed for that percentile. Interpolation between actual sample values is not used. For example, if 25 samples were ranked in ascending order, 10 percent of the samples would be 2.5 samples, and the concentration of the third value is shown in the 10th percentile column. Similarly, the 20th percentile would be the fifth value.

The site information, together with the summary results of the sampling activity, is intended to provide the data needed when making international comparisons on levels and trends of air pollution.

After a sufficient body of data exists from each site (data for at least 3 years), the frequency distributions can be used to test for trend. By plotting the data on probability or log-probability paper for each year, trends can be visually determined. By employing a slightly more sophisticated approach, using the frequency distribution, one could test for trends statistically by use of the Kolmogorov-Smirnov test (see any basic nonparametric statistics book).

PROBLEMS IN DATA SUBMISSION AND REDUCTION

There have been very few problems encountered thus far and those that have occurred can be categorized into two types. The first type can be expressed as *coding errors*. For example, one of the data sheets reported that the sampling interval was 4 hours when the data were hourly values. In this category are errors that reflect differences in handwriting ability and handwriting peculiarities among different countries. The one digit that seems to cause the most serious problem is zero (0), as it is often misinterpreted as a six. Coding problems, however, are usually caught, before they cause too serious a problem, through validating procedures, either by International Data Bank personnel or by the contributing country when they compare the results returned by the Data Bank with their own records.

The second type of problem is far more difficult to handle. It involves the assignment of parameter/method codes that are to be used in comparing results. Essentially, the judgment has to be made whether the method of collection or analysis is different from another "established" method's thereby causing a theoretical difference in the results; if so, a new code must be assigned. This precaution is necessary so that only results obtained by similar methods are compared with each other. For example, the Environmental Protection Agency had on file 33 different methods for determining ambient SO_2 levels; however, after our involvement with WHO, we have had to assign five additional codes to handle the different methods. If there were standardized methods of sampling and analysis (or equivalent), then this task would, of course, be unnessary; and the resulting data would be more reliable and more comparable.

ACKNOWLEDGEMENT

The authors wish to acknowledge Mr.Gerlald J.Nehls for his efforts in developing the *Data User's Guide*.

APPLICATION OF ADAPTIVE PATTERN CLASSIFICATION TO THE DERIVATION OF RELATIONSHIPS BETWEEN AIR QUALITY DATA

RONALD E.RUFF*

Environmental Protection Agency National Environmental Research Center Meteorology Laboratory, Research Triangle Park, North Carolina 27711, U.S.A.

SUMMARY

Adaptive pattern classification has been demonstrated as being a useful tool in some air pollution modeling problems. It is certainly not a panacea for all types of problems. It does, however, provide the air resources engineer with another valuable technique. This paper presents an example using one of the more basic pattern classification methods while alluding to the more complex approaches. These are treated in the references. [1,4]

INTRODUCTION

In the field of air pollution modeling, the general approach is to make restrictive assumptions concerning the formulation between variables. Unfortunately, it is usually the case that the available data are not extensive enough to prove or disprove the validity of these assumptions. The adaptive techniques presented in this paper avoid such restrictions by seeking a numerical solution between a set of input variables, called a pattern, and a categorized dependent variable. The computer models are adaptive since they are programmed to modify their internal structure as a function of some typical data, called a training set.

The approach outlined in the previous paragraph is commonly designated as an adaptive pattern classifier. The prime asset of this approach is its ability to solve for a highly non-linear relationship between a large set of variables. Such a feature is required in many air pollution modeling applications.

Section II presents an introductory discussion of adaptive pattern

---

* Researched described in this paper was conducted by the author at the University of Washington, Seattle under U.S.Public Health Service Grant T01-00029-07.

classification. A simple application of this theory is described in Section III. This involves an ozone prediction model developed from data made available by the San Francisco Bay Area Air Pollution Control District. The potential of adaptive pattern classification with respect to the data base to be generated by the Regional Air Pollution Study (RAPS) is treated in the final section. A brief description of the RAPS program is also presented.

ADAPTIVE PATTERN CLASSIFIERS

General Pattern Classification Problem

At its lowest level, the general pattern classification problem consists of techniques to separate patterns, such as groups of objects, sounds, odors, events or properties, into classes based on measurements made on the entities being classified. A pattern then is a set of observations which is used to describe one object, sound, odor, event, or property. Those patterns with similar properties are grouped together to form a class or category.

In the field of pattern classification, two different types of problems are encountered. One of these problems requires an approach in which the patterns can best be represented in linguistic or semantic form. In the second type of problem, the patterns can be described in numerical form as pattern vectors. It is this second type which serves as the foundation for this paper.

The numerical pattern classifier problem is illustrated in Fig. 1. The *preprocessor*, or *receptor*, senses the pattern in its

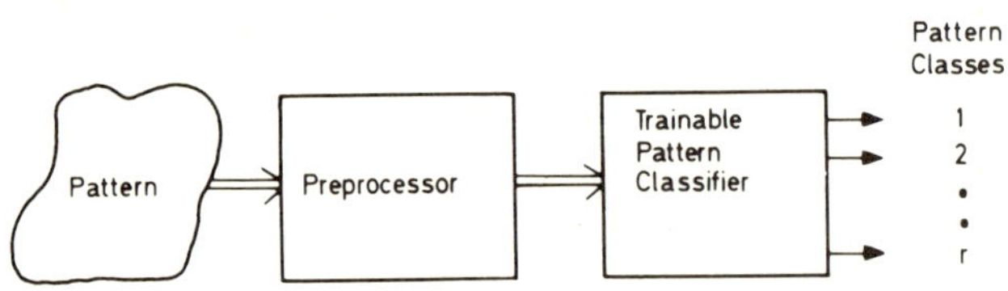

Fig. 1 - General Adaptive Pattern Classification Model

original form an generates a set of measurements or characteristics, in numerical form, corresponding to that pattern. The trainable (or adaptive) pattern classifier must then be able to classify these measurements according to the structure developed in its *learning* phase. A learning machine is defined by Nilsson [1] as any device whose actions are influenced by past experiences. This present work deals with the theory of a subclass of learning machines, those which can be *trained* to recognize or classify patterns.

A general model for the training phase of a pattern classifier is illustrated in Fig. 2. For the purpose of future discussion, all patterns will be assumed to be preprocessed. A set of patterns consisting of represen-

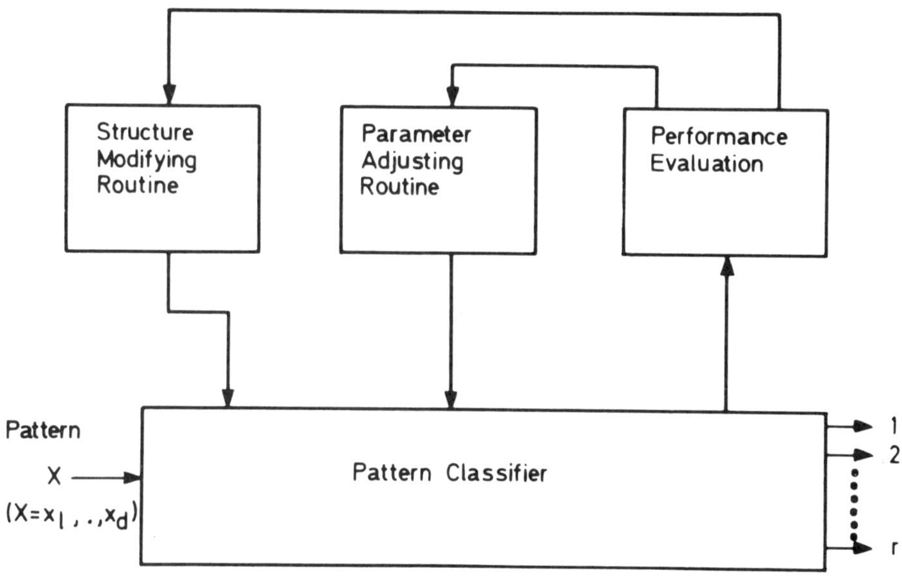

Fig. 2 - Training Phase of a Pattern Classification System

tative samples from each class, is used to train the classifier. These analog pattern vectors with known classifications are presented sequentially to the classifier. The performance is monitored by comparing the known classification with that computed by the classifier. Based upon success (or failure) in the classification, the classifier's parameters and/or structure are modified to improve the model with respect to the training set. This training process is iterative with a tendency to produce an incremental modification. Each cycle through the training set is called an *iteration*. Normally, several iterations are necessary before the training process converges.

Once the classifier is trained, it can be used (with fixes parameters

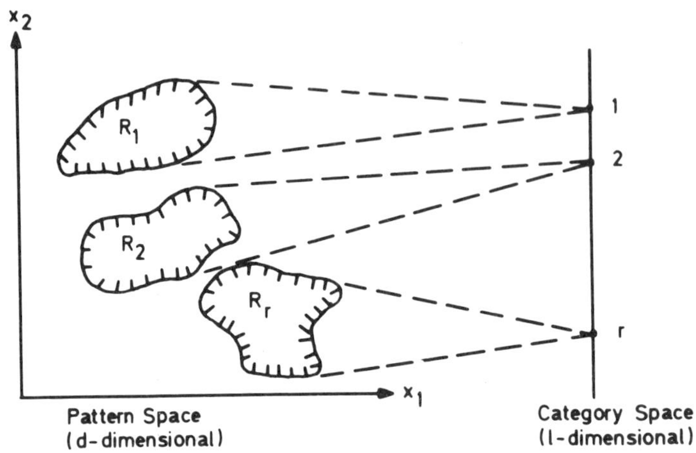

Fig. 3 - Two dimensioned pattern space example

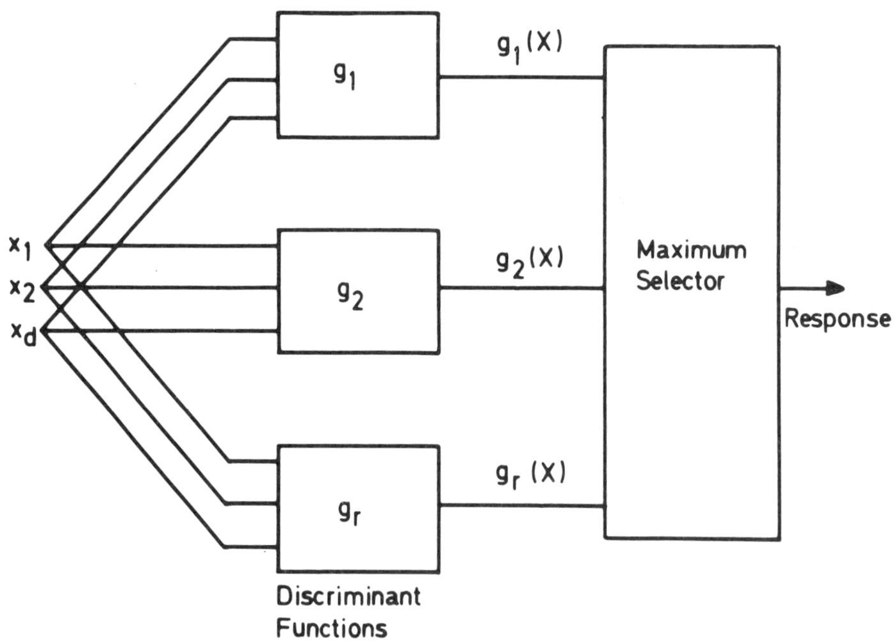

Fig. 4 - Basic discriminant function model

and structure) to classify patterns of unknown classification. However, it is more common to evaluate the performance first with respect to another set of patterns with known categories. The rationale is to obtain an error rate for a set of data which the classifier has not been optimized. This is the "acid test" that serves as a test with regard to the generality of the classifier.

Decision Surfaces and Discriminant Functions

The set of measurements that describe a pattern can be considered as a set of real numbers, x_1, x_2,..........x_d. Any pattern can be represented by a point in d-dimensional Euclidean space, E^d, called the *pattern space*. The rectangular coordinates of the point are the same numbers, x_1, x_2 ....x_d. The coordinates can also be treated as a vector, X, extending from the origin to that point. Both vector and point are denoted by X in future discussion.

The pattern classifier is then a device which maps the set of points or *regions* in E^d into the category numbers 1, ...,r. The points of different categories are separated by surfaces called *decision surfaces*. Fig. 3 pictorially represents this mapping process for two-dimensional space (d=2).

Decision surfaces can be implicitly defined by a set of functions containing r members. These *discriminant functions* are scalar, single-valued functions of the pattern X, termed $g_1(X)$, $g_2(X)$,....., $g_r(X)$. Refer to Fig. 4 for the general model of a pattern classifier implemented by discriminant functions. The number of discriminators,r, is equal to the dimension of category space. The maximum discriminator output for a particular pattern determines the category to which that pattern is assigned. More formally, the pattern is assigned to category i if:

$$g_i(X) > g_j(X) \quad \text{for } i,j = 1,2, \ldots\ldots..r \text{ and } j \neq i$$

Assuming the discriminant functions are continuous across decision surfaces, the equation for that surface separating continuous regions i and j is given by:

$$g_i(X) - g_j(X) = 0$$

The discriminant function approach is used exclusively in the models developed in this paper..

General Types of Training Methods

Training methods for learning machines are of two distinct types, *parametric* and *nonparametric*. The parametric appraoch is appropriate for classification tasks where each pattern category i,i=1,2,.....r, is known apriori to be characterized by a set of parameters, some of whose values are known. If the values of these parameters were known, adequate discriminant functions based upon their probability distributions can be specified directly. The pattern classifier, in this case, can assume a fixed structure. More commonly, the parameter values are not known. The parametric approach then consists of making assumptions about the form of the underlying probability distributions and then using the training patterns to estimate these distributions. The discriminant functions are subsequently determined by these estimates. Refer to Nilsson [1] and Fu [2] for a more comprehensive treatment of the parametric approach.

The non-parametric approach is more general since it does not make any restrictive assumptions about the probability distributions. Instead, functional forms are assumed for the discriminant functions such as linear, quadratic, and piecewise linear. These forms have unspecified coefficients which are adjusted in the training process in such a way that the discriminant functions perform well on the training set. Nagy [3] offers a more comprehensive comparison between the parametric and non-parametric techniques which is argumented by specific examples. This research concentrates on the more general non-parametric approach.

One final requirement, for both techniques, is that the training set must contain a representative group of patterns that are typical of those which ultimately must be classified.

Functional Forms of Discriminant Functions

Because of the simplicity involved, much of the work in the field of pattern recognition has concentrated on linear machines. There are two inherent problems with assuming a linear solution. First, there is no reason to believe "real world" problems will be linearly separable. The second problem is concerned with the selection of pattern features in the preprocessing stage. A poor choice of these features tends to complicate the solution, reducing the chance for a linear solution (if one exists). In spite of these problems, the linear machine, because of its simplicity, merits a "first look" in attacking most problems. When the solution does not meet expectations, more complicated forms are considered.

Linear Discriminant Functions

A linear machine can solve any problem where the categorizations are such that the corresponding pattern space can be partioned by linear decision surfaces. A two-dimensional example of that type of pattern space is illustrated in Fig. 3. Note that a straight line can be drawn separating any two of the pattern regions.

The general discriminant form for a linear machine is given by:

$$g_i(X) = w_{i1}.x_1+w_{i2}.x_2 + \ldots\ldots\ w_{id}.x_d +w_{i,d+1}$$

Another concept that is useful is physically describing this type of problem is that of *weight space*. The weight vector, W_i, with components, w_{i1}, w_{i2} . . . . . $w_{i,d+1}$, is located in a (d + 1) - dimensional space called a *weight space*. An *augumented pattern vector*, X^+, in (d + 1) - dimensional space is a vector with the first d components the same as X and the (d + 1) st component with a value of unity. The purpose of the extra component is to make the effective threshold between discriminant functions equal to zero. The rationale will be demonstrated more explicitly when considering the Minimum Euclidean Distance concept. In the training phase, the values of the weight vectors are determined.

Minimum Euclidean Distance

One established way of implementing discriminant functions involving dot products is based on the concept of Euclidean distance. The Euclidean distance between multi-dimensional vectors W_i and X is:

$$|W_i - X| = \sqrt{(W_i - X).(W_i - X)}$$

Associated with each weight vector W_i is a category number i, i=1,2,...r. A minimum Euclidean distance (MED) classifier, with respect to points in space defined by w_{i1}, w_{i2}..........,w_{id}, places each point X into that category which is associated with the nearest point. For any X, the quantities $|W_i-X|$, i=1,2,.....,r are calculated and X is placed into the category associated with the smallest.

The MED concept is illustrated graphically by the two dimensional example of Fig. 5. The minimum Euclidean distance between pattern X and those three points in the weight space is the same as the shortest vector distance, ED_1. Therefore, the pattern is placed in the category corresponding to that point in weight space (category 1 in this case).

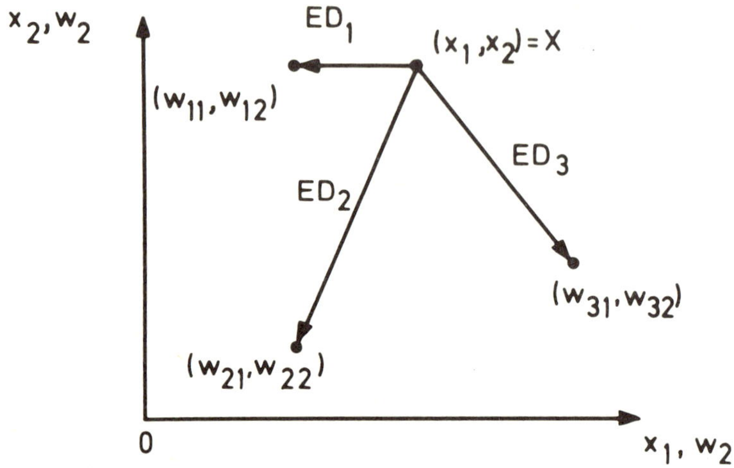

Fig. 5 - Two Dimensional MED Example

Iterative Training Procedures

During the training (learning) phase, both the pattern inputs and the corresponding correct categorization are used by the adaptive program to develop the model. *The technique essentially fits the model to the data.* The training phase relies on a performance evaluation for the model whereby the correct and predicted categorizations are compared for each pattern. This comparison is then used by the program as a basis subsequent adjustments to the model.

This paper considers only one of the many training procedures that exist. In the application presented in the next section, the MED discriminant function weights are iteratively adjusted by the program. These weight vectors can be thought of as centroids of pattern clusters. The classifier examines each new pattern, X, as it is introduced. Using the MED, the pattern is placed in the category corresponding to the closest weight vector. The program then examines the correct category for that pattern. If it is the same as preditected, the centroid is recalculated to reflect the features of the new pattern. If the predicted and correct categories are not the same, the centroid is shifted away from that pattern. This process is repeated until the solution converges. At this point, the adaptive pattern classifier is trained. It is emphasized that one cannot expect the prediction capability to be perfect (100%). The prediction accuracy is a function of the quality of data as well as the characteristics of the adaptive pattern classifier.

Summary

This section has discussed one of the more basic techniques in the field of pattern classification. It was assumed that the patterns were linearly separable. Techniques do exist to solve non-linear problems. A thesis by Ruff [4] successfully applies a non-linear technique to an air pollution modeling problem.

It is emphasized that this paper has only briefly treated one type of pattern classification technique. For a comprehensive overview of the field of pattern classification (or recognition) the reader is referred to the text by Nilsson [1].

A MODEL TO PREDICT VIOLATION OF THE CALIFORNIA STATE STANDARD FOR OZONE

Description

Ozone has been recognized as one of the components of photochemical smog. It is referred to as a secondary pollutant since it is not emitted directly in significant quantities by any known source. In the lower atmosphere, ozone can reach high concentrations because of the reactions involving other pollutants in the presence of sunlight. Various laboratory models have been developed in an attempt to explain the photo-oxidation phenomena observed in the atmosphere over urban areas. Because of the multiplicity of secondary species of pollutants, the mechanisms proposed so far are highly speculative.

The degree of photochemical smog cannot be explained on the basis of chemical reactions alone. Inversion levels and wind speeds are two of the meteorological variables which must be taken into account. They determine the degree to which the pollutants are removed over an urban area. Another important meteorological variable is the type of cloud cover. This affects the intensity of sunlight and, therefore, the entire photochemical process.

Due to the health hazard, it is desirable to have the ability to predict occurences of unusually high ozone concentrations. For this to be done, relationships between the ozone concentration and the controlling pollutant and meteorological factors must be developed. An early indication of a high ozone potential based upon these relationships can serve to initiate some reduction in source emissions or/and modification in the physical activity of the populace involved.

Formulation of an Ozone Predictor for San Jose

The California Air Resources Board adopted a State Standard for oxidants of 0.1 ppm for one hour. San Jose was chosed as the dependent variable since concentrations often exceed the State Standard during the summer and autumn months for which data is available. The terminologies "oxidants" and "ozone" are used interchangeably. In reality, the oxidant analyzer's output is composed primarily of ozone (over 90%). In predicting ozone concentrations, the approach consists of determining an ozone threshold related to health effects. The output of the predictor is consequently broken down into the following two categories:

(a) Category I - Ozone is less than 0.1 ppm. There is no potential health hazard

(b) Category 2 - Ozone equals or exceeds 0.1 ppm. There is a potential health hazard.

Ozone is described here and throughout the rest of the section in terms of the daily maximum hourly average.

The problem consequently reduces to one of predicting the correct binary output as a function of pollutant and meteorological variables. In San Jose, the State Standard for ozone was not violated before 12:00 noon (PDT). The maximum hourly average generally occurs between 1:00 pm (noon, PDT) and 2:00 pm. It is desired to predict the violation several hours in advance of its occurence. Therefore, the pollution inputs to the predictive model are measured at the time of prediction, several hours prior to normal occurence of the high ozone episode.

On the other hand, the meteorological inputs can be predicted for the most part. Therefore, the success of the ozone prediction generally relies strongly on a good prediction of the meteorological variables. It is emphasized at this point that the inputs to the predictive model are assumed to be related to the output. The model must be structured such that these assumptions can be contradicted by the predictor's performance. Consequently, if a contradiction is detected, the assumed input is discarded. This type of input modification is normally considered to be part of the preprocessing phase, as discussed in the former section.

While the model considers pollutant measurements from San Jose, corresponding measurements from Redwood City and Oakland are also fed into the predictor. Because of the prevailing winds and the peculiar topography of the area it is suspected that the pollutant transport, from other parts of the Bay Area, results in increased photochemical activity is San Jose.

Following is a list of trial inputs to be used in the San Jose ozone predictor:

(1) NO_2 - Selected because it reacts with sunlight to ultimately form O_3.

(2) CO - While there may be significant photochemical reactions involving CO, evidence indicates that CO is a good indicator for automotive exhaust pollutants which are known to play an important role in photochemical smog formation.

(3) $\Delta CO/\Delta t$ - The rate of change of CO in the atmosphere is indicative of the degree to which the primary pollutants are being dispersed.

(4) O_3 - The concentration of ozone in the early morning hours is indicative of the amount of photochemical activity.

(5) Percent sunshine (SS) - Used, along with temperature (T), to represent radiation intensity.

(6) Ventilation index (VI_1 and VI_2) - Computed at 0400 and 1600 hours daily on a scale of 1 to 5. Low indicies imply temperature inversions exist with low wind speeds, inhibiting dispersion. High indicies indicate that the atmosphere is donducive to a more thorough dispersion.

(7) Mean wind speed (MWS) - Is a measure of daily average surface wind speed.

The meteorological inputs represent averages of San Fransisco and Oakland data except for ventilation index which is calculated from the inversion base and wind speed at Oakland Airport. Pollution inputs corresponding to San Jose, Redwood City and Oakland are used in the model. The exception is the ozone input Only the San Jose ozone measurement is considered. It is not the purpose of this research to justify the use of any particular set of inputs. The object simply stated, is the optimization of a predictive model given a set of related inputs. Therefore, no further attempt is made to explain the rationale behind the variables involved.

The preceding paragraphs are summarized in the following mathematical formulation:

$$y_{t'} = f\left\{ MWS, T, SS, VI_1, VI_2 \left[(NO_2, CO)_{rc}, (O_3, NO_2, CO)_{sj}, (NO_2, CO)_{ok}, (\Delta CO)_{rc}, (\Delta CO)_{sj}, (\Delta CO)_{ok} \right]_t \right\} \quad (1)$$

where $y_{t'}$ is the categorized output over range of times the State Standard is normally violated; t corresponds to the time prediction is made; rc, sj, ok are subscripts denoting measurements in Redwood City, San Jose, and Oakland, respectively. Other variables are as defined in the previous paragraph. An adaptive pattern classifier, using MED discriminant functions as discussed in the former section can numerically solve the relationship designated by equation 1. If any of the dependent variables prove to be misleading or irrelevant, then a technique must be devised to allow for their deletion. The time of prediction t, is another variable that must be considired. If an accurate prediction is made in the early morning hours, such as 4 a.m., abatement action can be taken if necessary to reduce the amount of emissions. On the other hand, a later prediction, such as 9 a.m. is not as effective in curtailing the sources but can serve to warn the general public to modify their physical activity. The approach will be to optimize the predictor over a time period ranging from 2 a.m. to 10 a.m. in one hour increments. Therefore, the model is predicting from twelve to three hours in advance of the normal daily ozone peak.

One final note concerns the specific set of data made available for this research. The Bay Area Air Pollution Control District maintains continuous air monitoring stations in each of the counties under its jurisdiction. As of Semptember 1968, these stations were located at Port Chicago, Richmond, San Rafael, San Francisco, Redwood City, and San Jose. A seventh station, operated by the California State Health Department, is located in Oakland. Each station measures the following pollutants: carbon monoxide, nitric oxide, nitrogen dioxide, hydrocarbons, oxidants and particulate concentrations. Data for gaseous pollutants, as previously specified, covering the period of June 1, 1968 through December 31, 1968 was made available for this research. There are times, however, during this period when certain stations were inoperable. See Fig. 6 for a map of the Bay Area showing the location of the seven stations.

Since the degree of photochemical smog is strongly dependent upon time of year, the specific model is optimized over a limited time period of three months. Implied in this approach, is the fact that the time of year is itself one of the variables. In an attempt to hold this variable constant, the training set consists of August and October data. The model is subsequently evaluated for September data. One further restriction is that only week days are considered in the analysis. The rationale is that week days are generally characterized by similar source emission patterns. Conversely, weekend emissions are probably different from those during the week. Since the training set has been limited to two months, the actual number of patterns (training predictions) is limited to forty-five, assuming the pollution data is complete. While it is doubtful that a training set of this size can legimately represent all patterns typical of the general August to October period, the techniques involved are still the same. The generality of results in the model, however, is certainly subject to question.

Fig. 6 - Bay Area Map Showing Monitoring Station Locations

Implementation of an Ozone Predictor by an Adaptive Linear Classifier Augmented by Interactive Input Modifications

The objectives of the ozone predictive model, briefly restated are:

(a) Optimization of numerical relationship between output category and input variables.

(b) Elimination of irrelevant inputs.

(c) Optimization of model as a function of prediction time.

A function diagram of a model designed to consider all of these objectives is illustrated in Fig. 7. The particular objectives, to which each function is addressing itself, is indicated as such on the diagram.

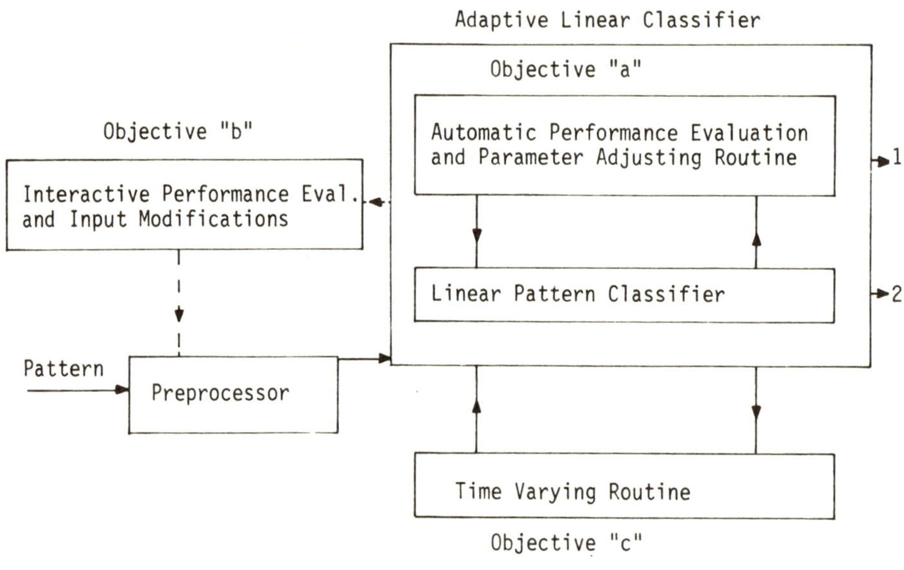

Fig. 7 - Ozone Predictive Model Function Diagram

A program was developed to implement the model of Fig. 7.

Description of the individual functions is given in the following subsections.

Adaptive Linear Pattern Classifier

The program employs training heuristics identical to those discussed in the former section. One small modification is the addition of a threshold as shown in Fig. 8. The use of a threshold is suggested by the role of sunlight in the formation of photochemical smog. Unless there is a certain amount of sunlight during the morning and early afternoon hours, the re-

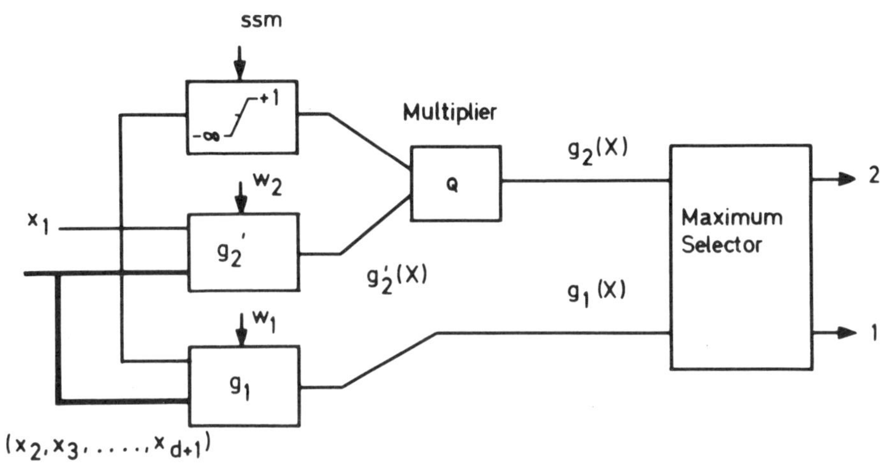

Fig. 8 - Linear Classifier with Threshold

sulting reduction in photochemical reactions will be significant. In this case, it is extremely unlikely that the ozone State Standard will be reached. Therefore, it is reasonable to model the effect of sunlight by the establishment of a sunshine threshold, SSM. The discriminant function for category 2 thus becomes:

$$g_2(X) = g_2'(X), \text{ if } x_1 \geqslant SSM$$

$$g_2(X) = -\infty, \text{ if } x_1 < SSM$$

where x_1 corresponds to the predicted percent sunshine for the day, previously referred to as SS. The routines establish a value for SSM during the first iteration by selecting the minimum value of SS correspondeing to an output in category 2.

In linear pattern classifiers, with similar inputs, the variables are usually weighted equally. In problems involving dissimilar types of inputs, the variables must be scaled as a means of compensation. Such is the case here where pollutant measurements are being compared with meteorological variables. The technique involves multiplying each meteorological input by a scale factor. The scale factors were chosen so that a given percentage change in each variable corresponds to about the same percentage change in the discriminant function. Ideally, the scale factors could be adaptively adjusted. However, the size of the traning set is too small in relation to the number of variables for such an adaptive process to work in this case.

Program Description

The program, OZOPT, was developed so that various combinations of input variables could be tried in rapid succession. Up to sixteen different combinations of input variables can be tried in one program application. This is implemented by use of a masking function which multiplies each of the input variables by zero or one. A zero multiplier dictated that the variable is to eliminated.

Since the OZOPT program was written for use with a batch processing computer (CDC 6400),it is not truly interactive in the sense that the programmer is communicating with the computer on a real time basis. However, by use of control cards, the interactive process can be realistically simulated. Each control card contains a trial masking function. The model adaptively optimizes the numerical relationships at a fixed time using the techniques as described in the previous subsection.

Once a good set of input variables is established, their performance at various times can be analyzed. Unfortunately, an input variable set, which is considered optimum for one time, is not necessarily optimum for all times. Optimizing the set of input variables at each time of interest requires reapplying in program OZOPT for each of these cases.

One could select the time corresponsing to the prediction of highest accuracy as the optimum prediction time. However, it is recognized that an early prediction is more effective in terms of source emission abatement and human preparedness. While recognizing that there are some "trade-offs" between prediction accuracy and amount of advance warning, these are not considered further in this research. The models hereinafter are judged on the basis of prediction accuracy.

The optimization of the ozone prediction model for San Jose consists of two general phases as follows:

(1) - The first phase involved interactively selecting a set of input parameters for a specified prediction time.

(2) - The second phase optimizes the numerical solution over a range of times given the set of input variables selected in the first phase.

Each phase is described in more detail in the following paragraphs.

The time of prediction must be selected to determine that part of the data to be used. In this example, it was arbitrarily set at 9 a.m.

Next, an initial set of masking functions is selected by means of control cards (one card per function). Up to 16 of these masking functions can be used in one application of the program. One possible approach is to try to determine the effect of each of the fifteen variables. This trial consists of optimizing the model for all inputs and then eliminating each of the fifteen other variables, one at a time.

Results of the first application of OZOPT indicate that no single pattern input variable exhibits a pronounced effect on the classification accuracy.

The next step involved eliminating groups of input variables. Since there are an extremely large number of possibilities, one might make such logical choices as eliminating all Oakland inputs or all NO_2 inputs. These types of trial input deletions were the basis for the second application of OZOPT. The best results correspond to the case where all NO_2 inputs are deleted. While the interactive process described in this subsection does not guarantee an optimum set of inputs, it does yield good results in the particular problem being considered after two applications. Now that an adequate set of inputs has been found, prediction accuracy versus time is next to be considered. For future discussion, it is assumed that the input set excludes the NO_2 inputs. The prediction distribution for this input set at 9 a.m. follows:

Predicted Category	Measured Category 1	2
1 (.1 ppm)	25	1
2 (.1 ppm)	4	10

Table 1 - Predicted distribution versus measured distribution

* Results are expressed in number of occurrences

Time Optimization with Fixed Inputs

Program OZOPT was applied to the problem with resulting prediction accuracies as shown in Fig. 9. For the sake of comparison, the accuracy of the "meteorological variables" only case is displayed also. The comparison suggests that the pollutant inputs do play a significant role in the type of predictive model under consideration.

It is emphasized that the prediction accuracies being discussed rely on accurate predictions of the meteorological variables. Forecasting variables such as ventilation index, percent sunshine, maximum daily temperature, and mean wind speed is itself a research area. Consequently, the accuracies mentioned here will decrease in reality as a result of the uncertainty of the meteorological variable predictions. On a comparative scale, it is reasonable to assume that the later prediction accuracies will not be diminished to the extent of the early ones as a result of errors in meteorological forecasts.

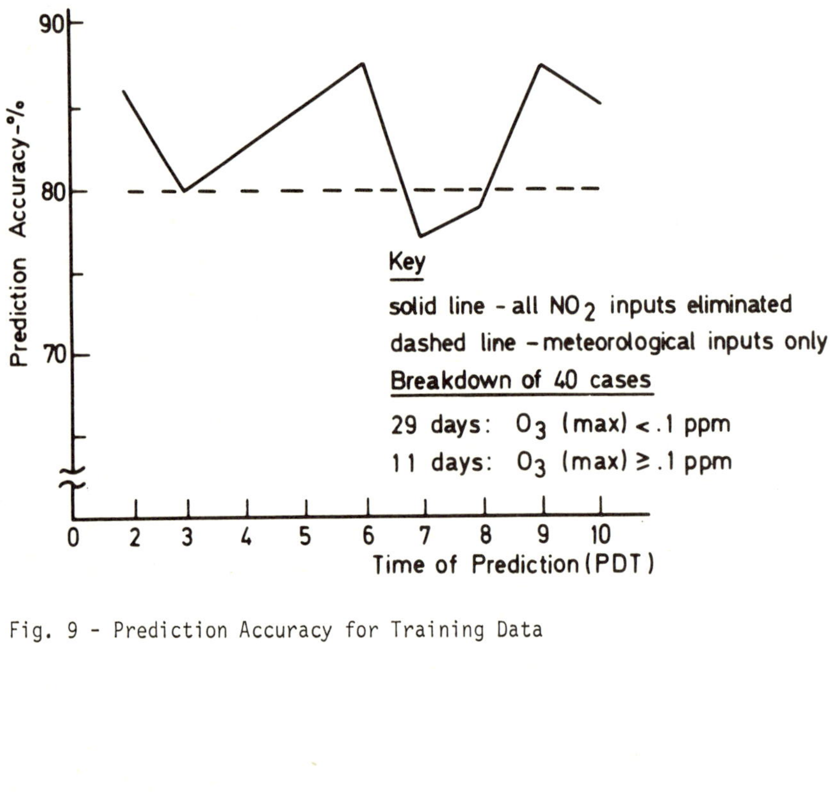

Fig. 9 - Prediction Accuracy for Training Data

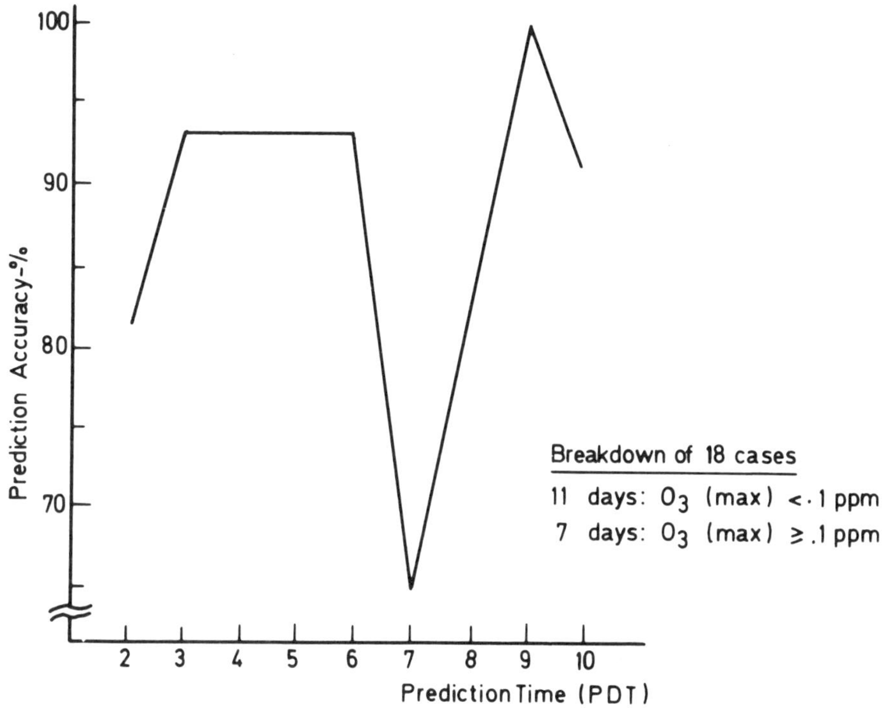

Fig. 10 - Prediction Accuracy for September Data

Application of Model to Independent Data

The predictive model now has a fixed structure determined by the utilization of the August/October training set. This model, using all inputs except those for NO_2, was tested for the September data set consisting of eighteen cases (days). Refer to Fig. 10 for the prediction accuracy versus time plots for September. As before, the 7 a.m. prediction was worst. This can partly be attributed to the high CO peaks corresponding to the morning rush hour. The model exihibits a tendency to be dominated by these high peaks. It was shown in Fig. 9 that a better set of input variables for this 7 a.m. prediction exists - elimination of all pollution inputs results in an accuracy of 80.0% for the training set. Aside from the 7 a.m. prediction, the model performs better on the independent data from September than on the training set.

One final result concerns the sunshine threshold. Using the value for SSM determined for the training set by program OZOPT, the patterns were classified correctly all six times in the September set for which the minimum was not exceeded.

Conclusions

This section has outlined a general appraoch to developing a model capable of predicting violation of state standard levels for ozone. Due to the small training set, the generality of the model with respect to San Jose is certainly questionable. However, its success is encouraging. Since the training set was small, more sophisticated training techniques were not attempted. The fixed scale factors which multiply the input variables are possibly too limiting. With a larger set of data, these could be adaptively optimized by the program.-

While emphasis is placed on the model's ability to predict, it should be emphasized that such a model can be used to establish cause-effect relationships. There were to interesting results that were encountered, aside from the prediction capability:

(1) The model performed better when the nitrogen dioxide inputs were eliminated. This contradicts the basis for all photochemical simulation models if one accepts the data as accurate. More likely, the NO_2 data were inaccurate enough to affect the calculation.

(2) The model performed better when data from the upwind cities were included in the input pattern. This suggests that the primary pollutants, emitted in the upwind cities, are being transported while reacting to affect the formation of ozone in San Jose.

This section discusses the application of an elementary adaptive pattern classifier to a small data base. Analysis of more comprehensive data bases most likely will require some of the more sophisticated techniques in the field of adaptive pattern classification. In particular, the extraction of the information and relationships in the proposed RAPS data base will require such techniques.

APPLICATION TO RAPS DATA BASE

The Regional Air Pollution Study (RAPS) is the most comprehensive research field program of its type ever undertaken by the United States Environmental Protection Agency. Field data from RAPS will be acquired over a period of several years from the St.Louis air quality region. The primary data source will be an automated network of twenty-five stations, connected via telephone lines to a central data acquisition facility. Following is a partial listing of parameters to be measured:

(1) Carbon monoxide
(2) Sulfur dioxide
(3) Ozone
(4) Nitrogen dioxide
(5) Nitric oxide
(6) Non-methane hydrocarbons
(7) Particulates
(8) Wind speed and direction
(9) Relative humidity
(10) Solar radiation
(11) Temperature

The automated network will be supplemented by field experiments and routine laboratory analysis. Part of the laboratory work will center on aerosol composition analysis. This will be routinely performed on the bag samples collected at the stations.

The field experiments will include a comprehensive emissions inventory acquisition. Temporal resolution to an hour will be ascertained for both mobile and stationary sources. Instrumented helicopters will routinely obtain vertical profiles of pollution and meteorological data and further be used to track air parcels in tracer experiments. Many other field experiments are planned during the course of RAPS. For an overview of the RAPS program, the reader is referred to the RAPS Study Plan [5].

The primary objective of the RAPS program is the development of mathematical simulation models of air pollution processes. The first step is the acquisition of a data base to support the model development. Current air quality simulation models are generally limited in the number of variables considered. This stands to reason since the data base available to validate these models has similary been limited in scope. As the data base becomes more extensive, techniques are needed to extract the additional information. Preliminary to the development of a dynamically based mathematical model, it is essential to ascertain what relationships are implied in the data base.

The rationale of the previous paragraph can be clarified by considering the status of the photochemical simulation models. At present, the photochemical models treat from ten to forty reactions, depending upon which specific model is being used. The models concentrate on gaseous constituents, completely ignoring the effect of atmospheric aerosols. This is logical since comprehensive aerosol data has been scarce. The acquisition of an aerosol data base is a vital ingredient of the RAPS. It is essential to evaluate the effects of aerosols on existing kinetic models. Before embarking on a massive model development program to incorporate aerosol effects into the photochemical model, it is more practical to examine the data to gain insight to the problem. A technique, such as adaptive pattern classification, can be used to determine the nature of the aerosol effect on the gaseous pollutants. This adaptive approach, as previously mentioned, is capable of solving for highly non-linear relationships between a large set of variables.

REFERENCES

1 N.J.NILSSON, *Learning Machines*, McGraw-Hill, New York, 1965

2 K.S.FU, *Sequential Methods in Pattern Recognition and Machine Learning*, Academic Press, 1968

3 G.NAGY, *IEEE Transactions on Audio and Electroacoustixs*, AA-16, no.2, June 1968

4 R.E.RUFF, *Computer Modelling of Urban Air Quality.Using Adaptive Pattern Classification*, doctoral dissertation, University of Washington, Seattle, 1971; University Microfilms, 300 Neeb Road, Ann Arbor, Michigan 48106

5 RAPS, *Series no.1 Study Plan, EPA-NERC*, Research Triangle Park, North Carolina, in publication

DATA MANAGEMENT FOR THE COMPUTATION OF URBAN POLLUTION MODELS

G.COCQUYT, J.F.DE GREEF, J.VANDERVEE

C.E.N./S.C.K., Mol, Belgium

SUMMARY

The basis of all pollution computation is a mathematical mode. For air pollution, the result at a point x, y due to pollution sources at $x_i y_i$ depends on a lot of meteorological and source variables such as the wind direction wind speed, temperature gradients, height of the mixing layer, characteristics of the stacks, temperature of the effluents etc..

In principle two methods can be used for the handling of these meteorological and source data. One method starts from a "plume-data bank" and avoids insertion of the data corresponding to the source coordinates, wind direction and pollutant flow; the other method computes" pollution charts" recalculating the pollution plumes for each source. The last method has the advantage of avoiding discretisation of all variables corresponding to the source characteristics.

Advantages and disadvantages are considered as a function of computer time and the objectives of the computation which may be, for example, the frequency distribution of hourly mean values due to many sources in an area of several km^2, the contribution of new sources to the existing background of a region, or the comparison between different sites.

INTRODUCTION

The computation problems encountered in the study of air pollution are so different and machine time consuming that it is worthwhile to think about data management in order to decide which is the best approach.

In this paper we will discuss two possible methods of data management. The first one starts with the calculation of pollution charts, given the characteristics of the sources such as their coordinates, their pollutant flow etc., and the weather characteristics such as the wind direction, wind speed, inversion height, etc..A flow chart for this method (Fig.1) and a more elaborate discussion is presented. The main characteristics of a second

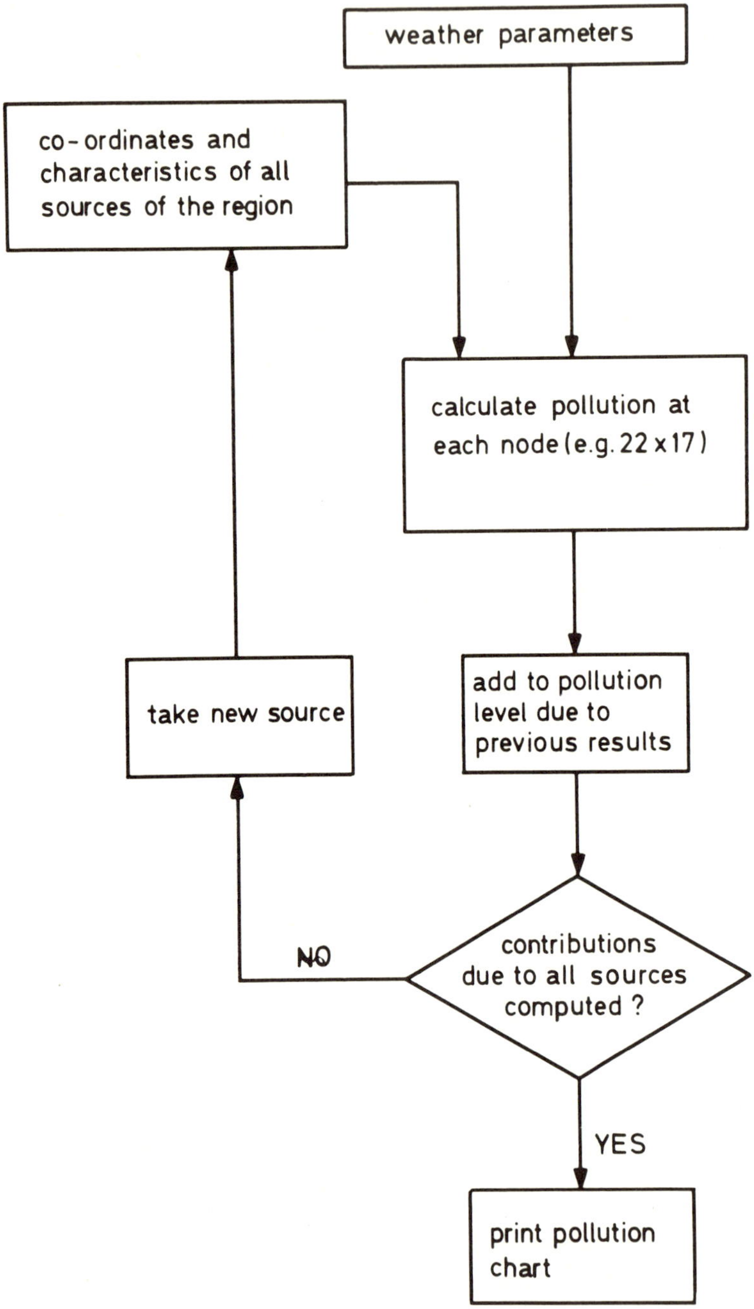

Fig. 1 - Scheme of computation of a pollution chart corresponding to one set of weather parameters

method which uses a data bank containing various pollution plume structures are given. The calculation of pollution charts, mean values etc. is performed using these plumes, shifting and rotating them corresponding to the coordinates of the source and the direction of the wind.

Finally, both methods are compared. Comparison is possible only by keeping in mind the aims of computation and the available means (data bank size and machine time). Therefore we will try to enumerate some frequently encountered problems.

1. Problems with stationary or moving sources. The sources may have changing coordinates or change their effluent flow Q in time: $Q(x_s, y_s, t)$.
2. Problems with stationary weather or with weather conditions that change with time. One type of problem in this class is the problem of nonstationary plumes. Here, only stationary plumes are considered and not the deformation of plumes due to transport in a time during which the weather characteristics (e.g. wind direction) change significantly to change the plume structure.

In both classes one may have to compute

a. A pollution chart
b. The pollution at one point as a function of time
c. Maxima, means and distribution functions
d. Comparison of meteorological data by the mean or by the distribution etc.

THE POLLUTION CHART METHOD

In this method one calculates the pollution at different points of a grid (e.g. a region of 22 x 17 km with a mesh of 1 km), due to each source separately. The scheme of computation is given in Fig. 1.

This scheme needs no explanation. Essentially for each grid point the pollution formula is computed as many times as there are sources. The loop in Fig. 1 is to be considered as an inner loop for Fig. 2 giving the computation scheme for all possible meteorological parameters.

Comparison of methods on the basis of computation time is only possible when the number of calculations can be estimated. An estimation of the computer time can be performed by calculating the number of times the inner loop (Fig. 1) has to be calculated. This number depends on:

1. The number of points (e.g. 17 x 22)
2. The number of sets of weather parameters

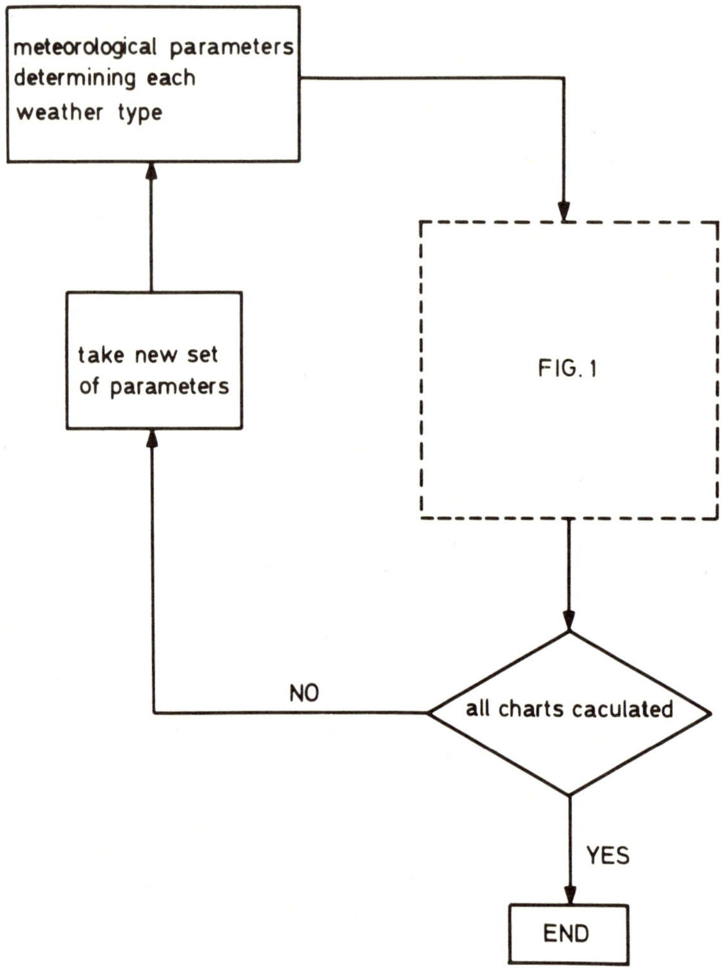

Fig. 2 - Scheme of computation of a set of pollution charts, each chart corresponding to one meteorological condition

These are A_{ϕ} = number of wind directions
A_u = number of wind speeds
A_e = number of stability categories
A_t = number of ambient temperatures
A_h = number of inversion layer heights

3. The number of sources

The product of all these numbers determines the computer time for the set up of the basic charts (results of Fig.2). These charts are to be used for the computation of problems as described in the introduction.

THE PLUME DATA BANK APPROACH

In this approach one calculates the plume structure for different source characteristics and meteorological conditions neglecting wind direction and source doordinates (in such a data bank the wind direction is fixed at $\phi = 0$ degrees, and the source coordinates are 0,0).

The mesh size for the plume structure must be chosen so as to allow linear interpolation in the calculation of the pollution, at an area grid point, due to the different sources, while keeping the size of the data bank within reasonable limits.

We will now try to determine the number of data necessary to describe a pollution plume, in the case of the well known Gaussian model [1]. In this model one assumes a Gaussian distribution perpendicular to the wind direction. The dispersion of this distribution depends on the weather stability category, which is determined by the ratio of the temperature gradient to the square of the wind speed, and the distance from the stack [2] .

In Fig.3 two pollution distributions are represented at different distances from the stack.

In principle the mesh size for the description of these distributions must be related to the dispersion σ . A description with 12 data covering 2σ corresponds more or less to Fig.3b. This number has been considered to be sufficient for our work. When we want to describe the narrow part of the plume (near the stack) as accurately as the broader part of the plume (away from the stack) we should use a mesh size that is related to the distance from the stack. However, this is difficult to do in practice because of the added programming problems and possibly increased machine time. Table 1 gives an idea of the variation of plume width as a function of the distance from the stack (x) and the stability categories (E_i i = 1 to 6). This table

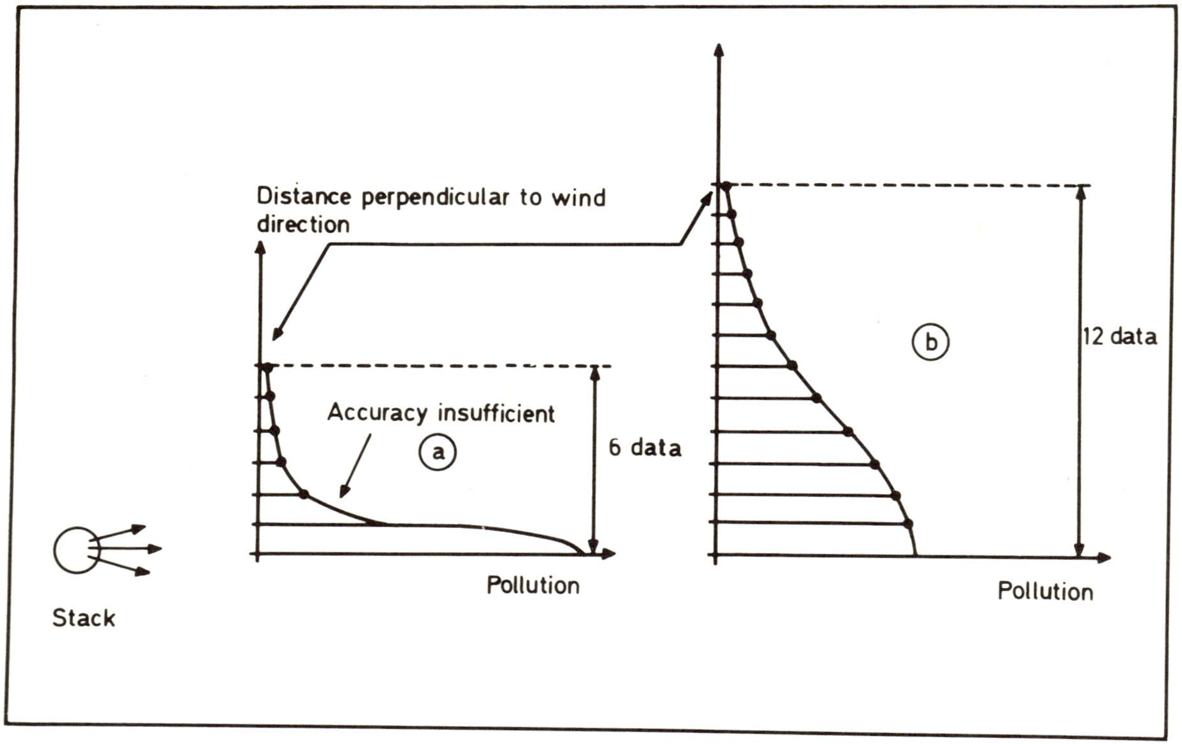

Fig. 3 - Pollution distribution at different distances from the stack

gives an idea of the mesh size to be used. Therefore the following principles must be observed.

Principle 1: It is not necessary to describe the detail the Gaussian distribution near the stack (h>30 m).

e.g.: the distribution with a σ of 33 m (i=1, x=500 m see Table I)need not to be described as the plume does not cause much pollution at ground level near the stack. It is enough to consider the plume structure at a distance where the pollution level is significant.

Principle 2: The maximum plume length is taken to be 20 km. Whenever it is deemed necessary to calculate pollution at longer distances, the size of the data bank must be increased proportionally.

Principle 3: It is sufficient to calculate the data for one half of the plume, due to symmetry. This principle is only to be applied perpendicular to the wind direction.

DETERMINATION OF THE MESH SIZE PERPENDICULAR TO THE WIND DIRECTION

With these three principles in mind, and using the formula for pollution given in ref. 1, we have found that one has to describe the plume with at least 12 points at a distance of $\pm$ 6 km. Since the plume is about 3 times as broad at 20 km we have estimated that at least 36 ponts are necessary for a good description of the plume distribution perpendicular to the wind direction.

DETERMINATION OF THE MESH SIZE IN THE WIND DIRECTION

In this case no reduction in data due to symmetry is possible. Here at least 24 data are necessary considering a variable mesh size. Again, due to programming difficulties one has to use a constant mesh. Even the principle of using only two mesh sizes, a smaller size for the critical part and a larger for the rest, is difficult to implement. The reason is that the critical part starts at widely varying distances for different weather conditions. This becomes clear considering e.g. the place of maximum pollution (Table II). Assuming a constant mesh per plume we have found that around 100 data are needed for the plume description along the wind direction.

DETERMINATION OF THE NUMBER OF DATA PER PLUME

The above considerations provide us with a figure of 3600 points for each plume in the data bank. It should be noted that the data bank gives information at ground level only. However, it is possible to adapt the source data (stack height) in order to calculate the pollution at other

Table I

COMPUTATION OF PLUME WIDTHS (1σ)

STAB.CAT. = 1 stable weather

STAB.CAT. = 6 unstable weather

X = Down wind distance

STAB CAT	= 1	= 2	= 3	= 4	= 5	= 6
X= 500	= 33.071	= 41.796	= 58.824	= 82.467	= 116.24	= 133.13
X= 1000	= 57.421	= 72.570	= 102.14	= 143.19	= 201.83	= 231.15
X= 1500	= 79.293	= 100.21	= 141.04	= 197.73	= 278.71	= 319.20
X= 2000	= 99.698	= 126.00	= 177.34	= 248.61	= 350.43	= 401.34
X= 2500	= 119.08	= 150.49	= 211.81	= 296.93	= 418.54	= 479.35
X= 3000	= 137.68	= 174.00	= 244.89	= 343.31	= 483.92	= 554.22
X= 3500	= 155.65	= 196.71	= 276.86	= 388.13	= 547.09	= 626.57
X= 4000	= 173.11	= 218.78	= 307.91	= 431.66	= 608.44	= 696.84
X= 4500	= 190.12	= 240.28	= 338.17	= 474.08	= 668.25	= 765.33
X= 5000	= 206.75	= 261.30	= 367.75	= 515.56	= 726.71	= 832.28
X= 5500	= 223.05	= 281.89	= 396.74	= 556.19	= 783.99	= 897.88
X= 6000	= 239.04	= 302.11	= 425.19	= 596.08	= 840.21	= 962.28
X= 6500	= 254.77	= 321.99	= 453.16	= 635.30	= 895.49	= 1025.6
X= 7000	= 270.25	= 341.55	= 480.70	= 673.90	= 949.90	= 1087.9
X= 7500	= 285.51	= 360.83	= 507.84	= 711.95	= 1003.5	= 1149.3
X= 8000	= 300.56	= 379.85	= 534.61	= 749.48	= 1056.4	= 1209.9
X= 8500	= 315.42	= 398.63	= 561.04	= 786.53	= 1108.7	= 1269.7
X= 9000	= 330.10	= 417.19	= 587.16	= 823.14	= 1160.3	= 1328.8
X= 9500	= 344.62	= 435.54	= 612.98	= 859.34	= 1211.3	= 1387.3
X= 10000	= 358.98	= 453.69	= 638.52	= 895.15	= 1261.8	= 1445.1
X= 10500	= 373.19	= 471.65	= 663.81	= 930.60	= 1311.7	= 1502.3
X= 11000	= 387.27	= 489.45	= 688.85	= 965.71	= 1361.2	= 1559.0
X= 11500	= 401.22	= 507.08	= 713.66	= 1000.5	= 1410.3	= 1615.1
X= 12000	= 415.05	= 524.55	= 738.25	= 1035.0	= 1458.9	= 1670.8
X= 12500	= 428.75	= 541.87	= 762.64	= 1069.2	= 1507.0	= 1726.0
X= 13000	= 442.35	= 559.06	= 786.82	= 1103.1	= 1554.8	= 1780.7
X= 13500	= 455.84	= 576.11	= 810.82	= 1136.7	= 1602.2	= 1835.0
X= 14000	= 469.23	= 593.03	= 834.63	= 1170.1	= 1649.3	= 1888.9
X= 14500	= 482.52	= 609.83	= 858.27	= 1203.2	= 1696.0	= 1942.4
X= 15000	= 495.72	= 626.51	= 881.75	= 1236.1	= 1742.4	= 1995.5

Table II

COMPUTATION OF DISTANCE FOR MAXIMUM POLLUTION

STAB.CAT. = 1 : stable weather

STAB.CAT. = 6 : unstable weather

X = effective plume height

STAB.CAT	= 1	= 2	= 3	= 4	= 5	= 6
X= 40	546	409	265	175	114	72
X= 80	1448	1084	703	463	301	189
X= 120	2560	1917	1243	818	532	335
X= 160	3837	2874	1862	1226	798	502
X= 200	5252	3933	2549	1678	1092	687
X= 240	6787	5083	3294	2168	1411	888
X= 280	8430	6313	4091	2693	1753	1103
X= 320	10172	7617	4937	3250	2115	1330
X= 360	12005	8990	5826	3835	2496	1570
X= 400	13922	10426	6757	4448	2895	1821
X= 440	15920	11921	7726	5086	3310	2082
X= 480	17992	13473	8732	5748	3741	2353
X= 520	20136	15079	9772	6433	4187	2633
X= 560	22348	16735	10846	7140	4647	2923
X= 600	24625	18441	11951	7867	5120	3221
X= 640	26965	20193	13086	8615	5607	3527
X= 680	29365	21990	14251	9382	6106	3840
X= 720	31823	23831	15444	10167	6617	4162
X= 760	34338	25714	16664	10970	7140	4491
X= 800	36906	27638	17911	11791	7674	4827
X= 840	39528	29601	19183	12629	8219	5169
X= 880	42201	31602	20480	13482	8775	5519
X= 920	44923	33641	21802	14352	9341	5875
X= 960	47694	35716	23146	15238	9917	6237
X= 1000	50513	37827	24514	16138	10503	6606

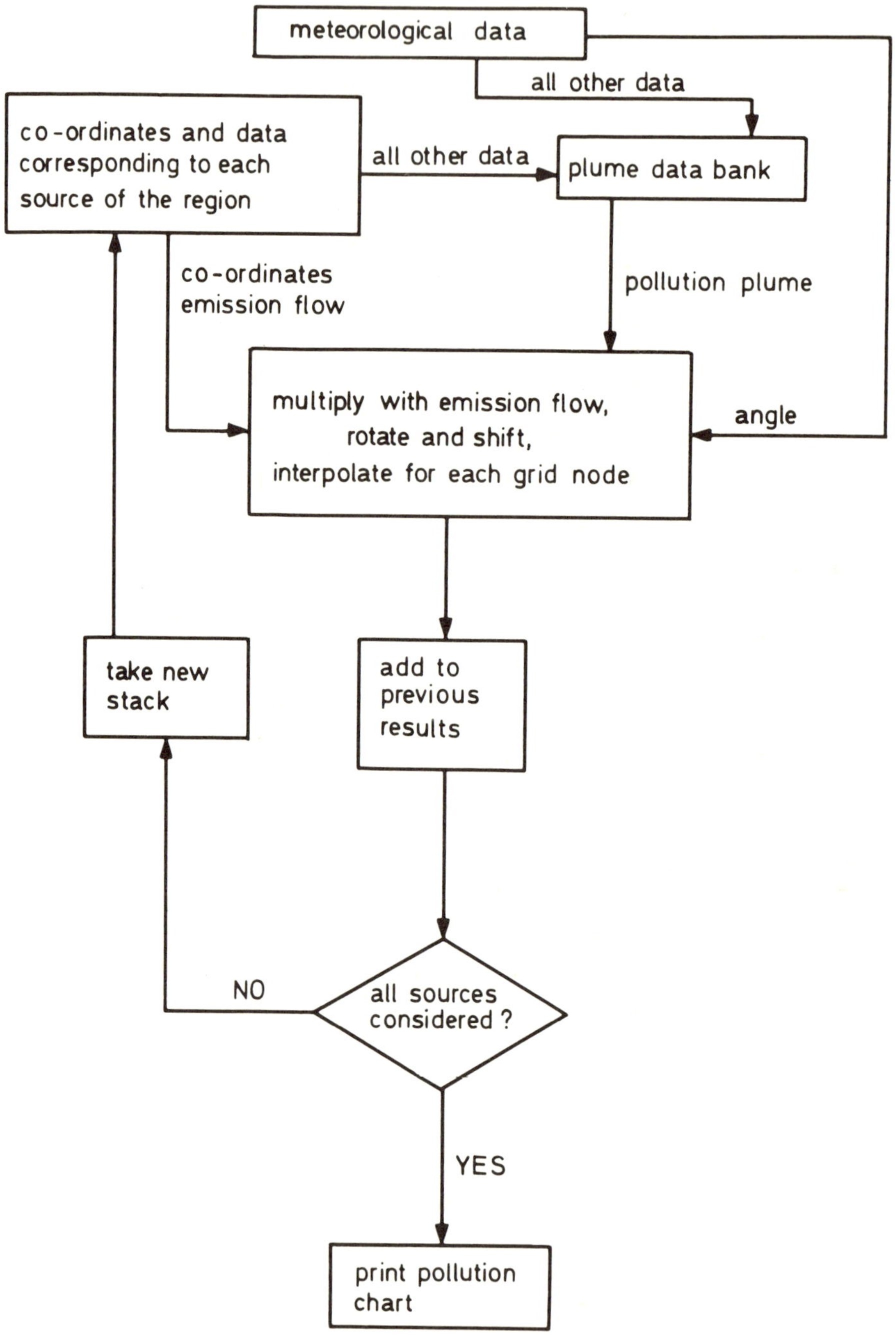

Fig. 4 - Calculation of a pollution chart for one meteorological condition using a plume data bank

levels provided the wind speeds to be taken into account do not depend on the stack height.

Fig. 4 shows the flow chart of the calculation of a pollution chart using a data bank.

FORMULA FOR THE DETERMINATION OF THE SIZE OF THE PLUME DATA BANK

The number of data necessary to perform the calculations, this is the number of data in the plume data bank per plume (36 x 100), multiplied with the number of plumes.

This is: $A_t.A_u.A_e.A_h.A_{gh}.A_q.A_d.36 \times 100$

The last part of this expression is discussed above

A_t, A_e, A_u and A_h are defined above

The remaining symbols are:

A_{gh} = number of geometrical stack heights

A_q = heath flows

A_d = number diameters

The last numbers can only be evaluated after sufficient experience, later we will estimate them in order to be able to draw conclusions.

DETERMINATION OF THE NUMBER OF POLLUTION CHARTS (FIRST METHOD)

Neither the mesh size of the pollution charts nor their number was discussed before. Both are problem dependent and need more detailed analysis. The problem which we will discuss here is the calculation of pollution at a particular point of the grid; the calculation of the moving average per hour or the moving average per 24 hours; the calculation of mean values over a given time interval (e.g. from 8 h to 18 h), and finally the corresponding distributions. The various means and distributions are calculated to check the model against experimental data and for eventual determination of stack heights or pollutant output rates. For this we assume the existence of a meteorological data bank containing all necessary weather data as hourly mean values of wind speed, wind direction etc., for e.g. one year.

THE PROBLEM OF DISCRETISATION OF THE WIND DIRECTIONS

A pollution chart is calculated for well determined weather conditions. Each conditions is given by 5 data: the wind direction $\emptyset$, the wind speed u, the stability category E_i, the inversion height layer H_{inv}. and the temperature T . These data together with the complete set of source characteristics.

We now concentrate our attention on the wind direction ϕ. It is clear that the discretisation of ϕ must be done, keeping in view covering the angle of the pollution plume (see Fig. 5) or the angle α under which the plume (1σ) at a certain distance can be seen.

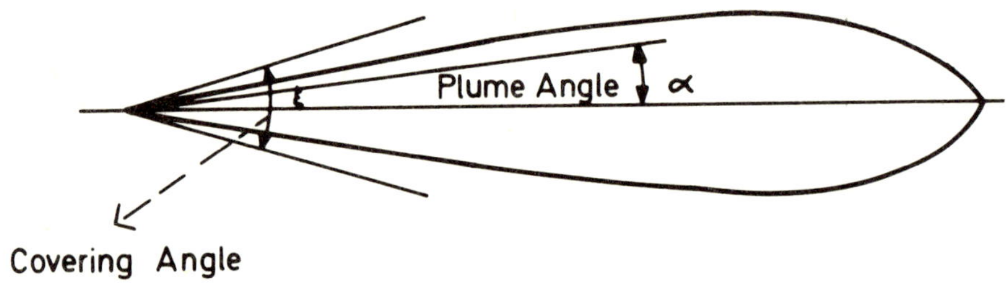

Fig. 5 - Covering angle and plume angle as calculated in Table III

For example, a discretisation by 30 degrees has no sense if the plume is contained in an angle of a few degrees since this results in a concentration of pollution over 30 degrees along the directions of discretisation. This can be demonstrated by computer results. A pollution chart looks like Fig. 6. This is quite a normal chart. It present the calculation corresponding to one set of weather parameters and a set of polluting sources, distributed ones as well as point sources, more or less arbitrarily spread over the region. Such a pollution space distribution exists as long as the weather parameters remain constant. Normally it exists only a relatively short time. Charts like this are to be superimposed according to their frequency of occurence to obtain e.g. the yearly mean chart. If the discretisation is too rough then a chart like Fig. 7 results.

This figure is obtained with a discretisation of 20 degrees giving 95% contours. The fact that we can observe a "star formation" tells us that the plume angles "in the mean" (over 1 year) are much smaller than the discretisation angle.

A better result is obtained if we use a discretisation of 10 degrees and still better with 5 degrees. A disadvantage however, is that the number of data corresponding to a smaller discretisation angle increase with the inverse of this angle.

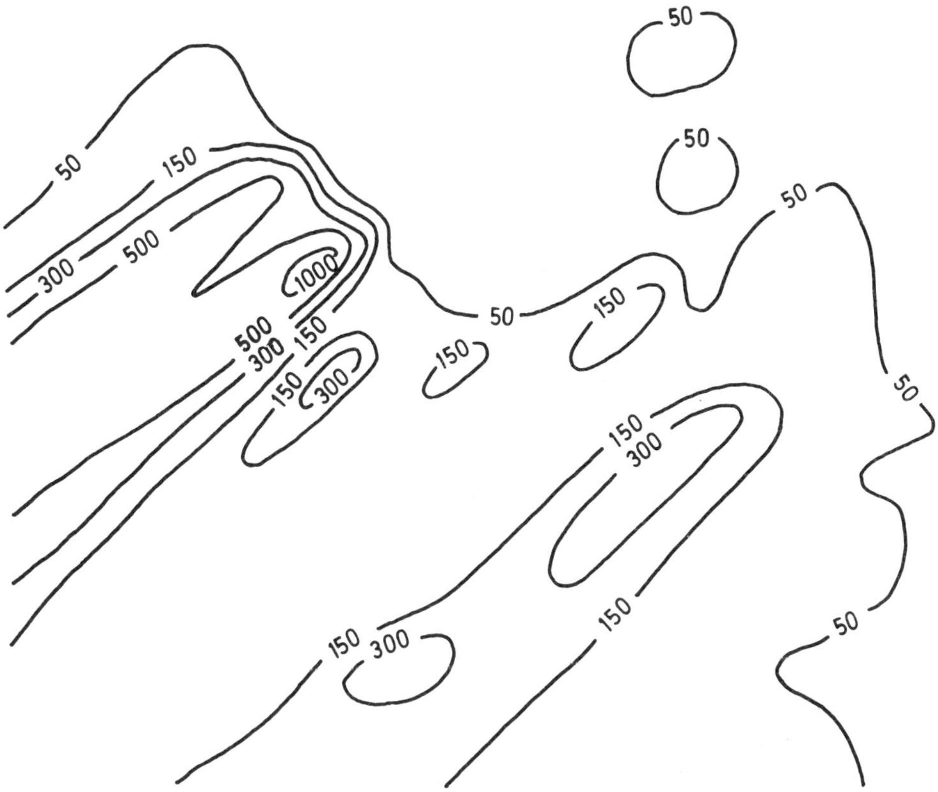

Fig. 6 - Example of a pollution chart result of the air quality display model developed for urban area by Mr.Legrand [4]

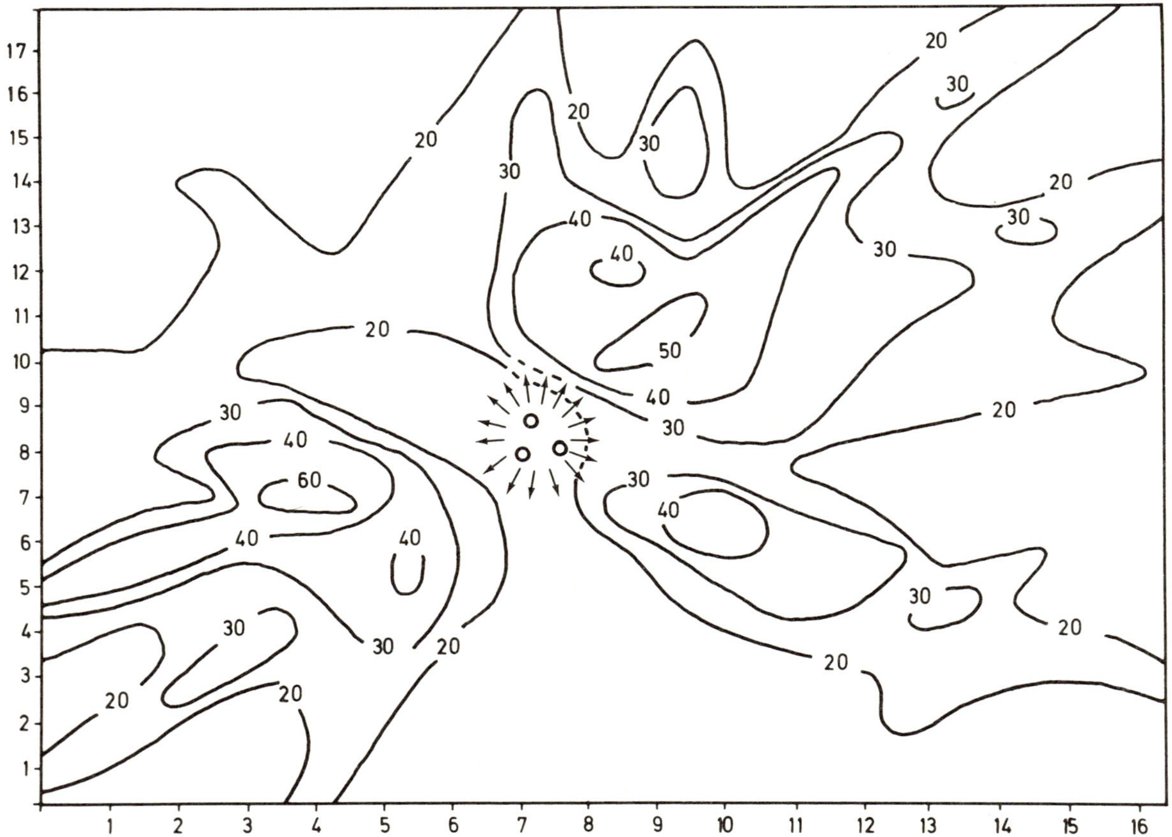

Fig. 7 - Result of calculation of code IFDM calculating the frequency distributions (95% values). An example of "Star formation" due to discretisation (20°)

Fig.8, 9,10 and 11 illustrate the effect of various discretisation angles. In these figures two Gaussion distributions with mean values separated by a different distance ($1.\sigma$, $2.\sigma$, $2.5\,\sigma$ and $3.\sigma$) are superimposed. We can observe that, only for a distance of 3σ the dip becomes severely pronounced. The corresponding discretisation may lead to star formation (Fig. 11).

It is therefore of interest to have an idea about the plume angle. This has been calculated and the result can be found in Table III .

In this table it can be seen that, for stability categories 1 and 2, the plume angle is about 2.5 degrees. Moreover these categories are most frequently encountered [4] and this explains the star formation in Fig.7.

This table together with the discussion permits us to decide that discretisation should be done at less than 10 degrees. A problem at the moment is that no meteorological data are available at less than 10 degrees of discretisation.

However, this problem of star formation with a discretisation of 10 degrees decreases if there is a sufficient number of stack not too far each other. Indeed, Fig.8,9,10 and 11 apply also to the superposition of two plumes existing at the same moment coming from two stacks 1,2, 2.5 and 3 dispersions apart. This means that in the case of several more or less arbitrarily distributed sources, the discretisation by 10 degrees may be sufficient.

REMARKS

To avoid star formation, two methods can be followed:

1. Discretisation by 10 degrees
2. Averaging of the Gaussian distribution within the discretisation angle. This method introduces an increase of computation time but may correspond more to reality since the wind direction does not remain stable for a long time.

DATA MANAGEMENT FOR THE COMPUTATION OF URBAN POLLUTION

The use of a plume data bank for direct calculation of pollution distributions for example. is drastically different from the use of a pollution chart.

In Fig. 4 is given a flow chart for the calculation of a pollution chart. It is clear that computation consumes less machine time than the method

```
VALUE OF K: 1

X= 210;Y= 0.0002-*
X= 220;Y= 0.0004-*
X= 230;Y= 0.0008-*
X= 240;Y= 0.0014---*
X= 250;Y= 0.0023-----*
X= 260;Y= 0.0036--------*
X= 270;Y= 0.0053--------------*
X= 280;Y= 0.0074-------------------*
X= 290;Y= 0.0098-------------------------*
X= 300;Y= 0.0122--------------------------------*
X= 310;Y= 0.0146--------------------------------------*
X= 320;Y= 0.0164-------------------------------------------*
X= 330;Y= 0.0177----------------------------------------------*
X= 340;Y= 0.0181-----------------------------------------------*
X= 350;Y= 0.0176---------------------------------------------*
X= 360;Y= 0.0163------------------------------------------*
X= 370;Y= 0.0143-------------------------------------*
X= 380;Y= 0.0120-------------------------------*
X= 390;Y= 0.0095------------------------*
X= 400;Y= 0.0071------------------*
X= 410;Y= 0.0051-------------*
X= 420;Y= 0.0034--------*
X= 430;Y= 0.0022-----*
X= 440;Y= 0.0013---*
X= 450;Y= 0.0007-*
X= 460;Y= 0.0004-*
X= 470;Y= 0.0002-*
X= 480;Y= 0.0001-*
X= 490;Y= 0.0000-*
```

Fig. 8 - Plotting of the sum of two Gaussian distributions K dispersions apart

```
VALUE OF K:2


X= 210;Y= 0.0002-*
X= 220;Y= 0.0004-*
X= 230;Y= 0.0007-*
X= 240;Y= 0.0013---*
X= 250;Y= 0.0021-----*
X= 260;Y= 0.0032-------*
X= 270;Y= 0.0046-----------*
X= 280;Y= 0.0062---------------*
X= 290;Y= 0.0078-------------------*
X= 300;Y= 0.0094-----------------------*
X= 310;Y= 0.0107--------------------------*
X= 320;Y= 0.0116-----------------------------*
X= 330;Y= 0.0121------------------------------*
X= 340;Y= 0.0124------------------------------*
X= 350;Y= 0.0124-------------------------------*
X= 360;Y= 0.0124-------------------------------*
X= 370;Y= 0.0124-------------------------------*
X= 380;Y= 0.0123------------------------------*
X= 390;Y= 0.0121------------------------------*
X= 400;Y= 0.0115----------------------------*
X= 410;Y= 0.0105--------------------------*
X= 420;Y= 0.0091----------------------*
X= 430;Y= 0.0075------------------*
X= 440;Y= 0.0058--------------*
X= 450;Y= 0.0043----------*
X= 460;Y= 0.0029-------*
X= 470;Y= 0.0019----*
X= 480;Y= 0.0011--*
X= 490;Y= 0.0006-*
X= 500;Y= 0.0003-*
X= 510;Y= 0.0002-*
```

Fig. 9 - Plotting of the sum of two Gaussian distributions K dispersions apart

```
VALUE OF K:2.5
X=  170;Y=  0.0000-*
X=  180;Y=  0.0000-*
X=  190;Y=  0.0000-*
X=  200;Y=  0.0001-*
X=  210;Y=  0.0002-*
X=  220;Y=  0.0004-*
X=  230;Y=  0.0007-*
X=  240;Y=  0.0013---*
X=  250;Y=  0.0021-----*
X=  260;Y=  0.0031-------*
X=  270;Y=  0.0045-----------*
X=  280;Y=  0.0061---------------*
X=  290;Y=  0.0077-------------------*
X=  300;Y=  0.0091----------------------*
X=  310;Y=  0.0101-------------------------*
X=  320;Y=  0.0107--------------------------*
X=  330;Y=  0.0107--------------------------*
X=  340;Y=  0.0104-------------------------*
X=  350;Y=  0.0099------------------------*
X=  360;Y=  0.0095-----------------------*
X=  370;Y=  0.0094-----------------------*
X=  380;Y=  0.0096-----------------------*
X=  390;Y=  0.0100-------------------------*
X=  400;Y=  0.0105--------------------------*
X=  410;Y=  0.0108--------------------------*
X=  420;Y=  0.0106--------------------------*
X=  430;Y=  0.0099------------------------*
X=  440;Y=  0.0088---------------------*
X=  450;Y=  0.0073------------------*
X=  460;Y=  0.0057--------------*
X=  470;Y=  0.0042----------*
X=  480;Y=  0.0028-------*
X=  490;Y=  0.0018----*
X=  500;Y=  0.0011--*
X=  510;Y=  0.0006-*
X=  520;Y=  0.0003-*
X=  530;Y=  0.0002-*
X=  540;Y=  0.0001-*
X=  550;Y=  0.0000-*
X=  560;Y=  0.0000-*
X=  570;Y=  0.0000-*
X=  580;Y=  0.0000-*
X=  590;Y=  0.0000-*
X=  600;Y=  0.0000-*
X=  610;Y=  0.0000-*
X=  620;Y=  0.0000-*
```

Fig. 10 - Plotting of the sum of two Gaussian distributions K dispersions apart

```
VALUE OF K:3

X= 210;Y= 0.0002-*
X= 220;Y= 0.0004-*
X= 230;Y= 0.0007-*
X= 240;Y= 0.0013---*
X= 250;Y= 0.0021-----*
X= 260;Y= 0.0031-------*
X= 270;Y= 0.0045-----------*
X= 280;Y= 0.0061---------------*
X= 290;Y= 0.0076-------------------*
X= 300;Y= 0.0090----------------------*
X= 310;Y= 0.0100------------------------*
X= 320;Y= 0.0104-------------------------*
X= 330;Y= 0.0101-------------------------*
X= 340;Y= 0.0094-----------------------*
X= 350;Y= 0.0085---------------------*
X= 360;Y= 0.0075------------------*
X= 370;Y= 0.0068-----------------*
X= 380;Y= 0.0067----------------*
X= 390;Y= 0.0070-----------------*
X= 400;Y= 0.0078-------------------*
X= 410;Y= 0.0088---------------------*
X= 420;Y= 0.0097------------------------*
X= 430;Y= 0.0103-------------------------*
X= 440;Y= 0.0103-------------------------*
X= 450;Y= 0.0097------------------------*
X= 460;Y= 0.0086---------------------*
X= 470;Y= 0.0072-----------------*
X= 480;Y= 0.0056-------------*
X= 490;Y= 0.0041----------*
X= 500;Y= 0.0028------*
X= 510;Y= 0.0018----*
X= 520;Y= 0.0011--*
X= 530;Y= 0.0006-*
X= 540;Y= 0.0003-*
X= 550;Y= 0.0002-*
X= 560;Y= 0.0001-*
X= 570;Y= 0.0000-*
X= 580;Y= 0.0000-*
```

Fig. 11 - Plotting of the sum of two Gaussian distributions
K disperions apart

Table III

COMPUTATION OF PLUME ANGLE IN DEGREES

STAB CAT	= 1	= 2	= 3	= 4	= 5	= 6
X= 500	= 3.7842	= 4.7784	= 6.7099	= 9.3656	= 13.088	= 14.910
X= 1000	= 3.2864	= 4.1507	= 5.8317	= 8.1485	= 11.411	= 13.015
X= 1500	= 3.0260	= 3.8222	= 5.3716	= 7.5093	= 10.526	= 12.013
X= 2000	= 2.8538	= 3.6049	= 5.0670	= 7.0858	= 9.9382	= 11.347
X= 2500	= 2.7270	= 3.4449	= 4.8426	= 6.7735	= 9.5042	= 10.854
X= 3000	= 2.6276	= 3.3194	= 4.6666	= 6.5283	= 9.1631	= 10.467
X= 3500	= 2.5463	= 3.2169	= 4.5228	= 6.3279	= 8.8841	= 10.150
X= 4000	= 2.4780	= 3.1306	= 4.4017	= 6.1592	= 8.6490	= 9.8823
X= 4500	= 2.4192	= 3.0564	= 4.2976	= 6.0140	= 8.4467	= 9.6521
X= 5000	= 2.3678	= 2.9915	= 4.2066	= 5.8871	= 8.2695	= 9.4506
X= 5500	= 2.3223	= 2.9340	= 4.1258	= 5.7745	= 8.1125	= 9.2718
X= 6000	= 2.2815	= 2.8825	= 4.0535	= 5.6735	= 7.9716	= 9.1115
X= 6500	= 2.2446	= 2.8359	= 3.9881	= 5.5822	= 7.8441	= 8.9663
X= 7000	= 2.2109	= 2.7934	= 3.9284	= 5.4990	= 7.7278	= 8.8339
X= 7500	= 2.1801	= 2.7544	= 3.8737	= 5.4226	= 7.6211	= 8.7123
X= 8000	= 2.1516	= 2.7185	= 3.8232	= 5.3521	= 7.5226	= 8.6001
X= 8500	= 2.1252	= 2.6851	= 3.7763	= 5.2867	= 7.4311	= 8.4960
X= 9000	= 2.1005	= 2.6540	= 3.7327	= 5.2257	= 7.3459	= 8.3989
X= 9500	= 2.0775	= 2.6249	= 3.6918	= 5.1687	= 7.2662	= 8.3080
X= 10000	= 2.0559	= 2.5977	= 3.6535	= 5.1152	= 7.1914	= 8.2228
X= 10500	= 2.0356	= 2.5720	= 3.6174	= 5.0648	= 7.1209	= 8.1424
X= 11000	= 2.0164	= 2.5477	= 3.5833	= 5.0172	= 7.0543	= 8.0665
X= 11500	= 1.9982	= 2.5247	= 3.5511	= 4.9722	= 6.9913	= 7.9947
X= 12000	= 1.9809	= 2.5029	= 3.5205	= 4.9294	= 6.9314	= 7.9264
X= 12500	= 1.9645	= 2.4822	= 3.4913	= 4.8887	= 6.8745	= 7.8615
X= 13000	= 1.9489	= 2.4625	= 3.4636	= 4.8499	= 6.8202	= 7.7996
X= 13500	= 1.9339	= 2.4436	= 3.4371	= 4.8129	= 6.7684	= 7.7405
X= 14000	= 1.9196	= 2.4255	= 3.4117	= 4.7775	= 6.7189	= 7.6840
X= 14500	= 1.9060	= 2.4083	= 3.3875	= 4.7436	= 6.6714	= 7.6299
X= 15000	= 1.8928	= 2.3917	= 3.3642	= 4.7110	= 6.6258	= 7.5779
MEAN	= 2.3154	= 2.9252	= 4.1131	= 5.7556	= 8.0830	= 9.2362*

followed in Fig.1. Indeed, the repeated calculations of the pollution formula take more time than shifting and interpolating the data taken from a data bank.

Whenever distributions are to be calculated it is better not to interpolate at all and to select the precalculated pollution data from the pollution charts. Fig.12 illustrates the two methods together. In both these methods the pollution at any preselected point can be computed as a function of time. For pollution distribution computations one has to handle the hourly meteorological data e.g. per hour or per day or for a whole year. This mean many shifts, rotations and interpolations when the output X(t) is calculated directly (Plume data bank method, solid lines). In the pollution chart method (broken lines) one just collects precalculated data which takes less machine time. Pollution charts are calculated once for ever for a well-determined region (a well-defined set of pollution sources).

These charts take into account the physical data and emission rate for each stack.

When these data change in time one has to update the whole set of pollution charts ($\pm$ 15 min for $\pm$ 900 charts* of 17 x 22 points): this is time consuming and cannot be done continuously. In such cases the plume data bank method is easier to use, though the machine time required is enormous e.g. for stack data which do not change with time, the pollution chart method took $\pm$ 4.5 hours for the calculation of distribution functions 100%, 97,5%, 95% ... in the form of 10 charts for 17 x 22 points.

Fig. 13 gives the result of such a calculation. At first glance the separation into two parts is more or less surprising.

Superposition of the result with the distribution of the wind directions (Fig. 14) explains the fact that the chart is split into two parts, the wind direction probabilities are higher at the corresponding angles.

The calculations are performed with the pollution chart method, using a discretisation of 10 degrees. One may observe that the start formation has disappeared. We have used the same method with meteorological data which contained wind directions every 20^{0} and we could observe a significant star formation (Fig. 7).

Remark: In Fig. 12, "mean emission only" is written for the data that comes from the block "Emission per stack as a function of time Q(t)". This is in general the case, when one has to be calculate distributions in a region due to many stacks. When one of them is time-dependent, one has to

* product $A_q A_u A_i$

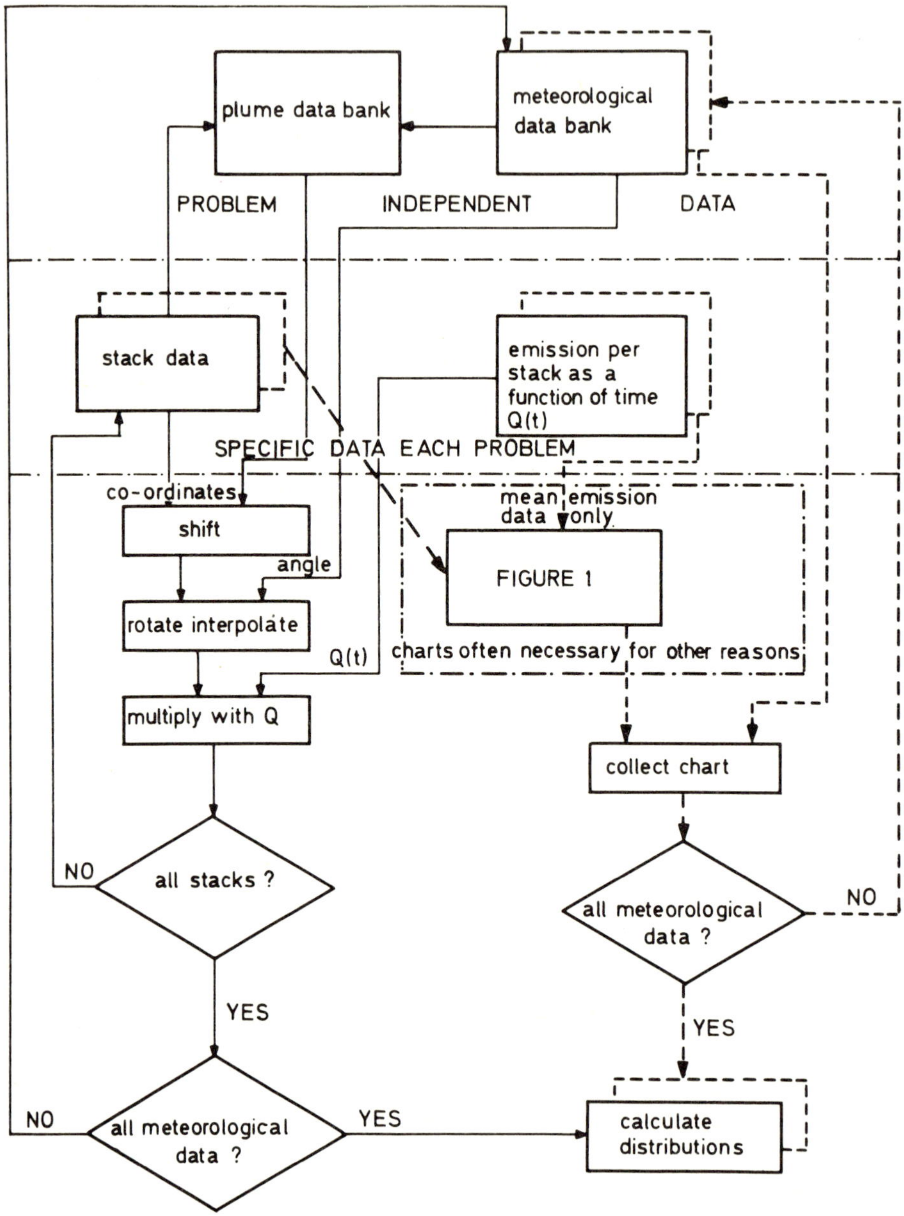

Fig. 12 - Summary of both methods
Solid lines: plume data bank
Broken lines: charts first

update the charts, which is time consuming. However, when one wishes to know the distribution due to one time-dependent stack only (or several that change propţionally), updating is not time consuming since it is performed by simple multiplication of all collected data (the chart) by a factor that is time-dependent.

THE AMOUNT OF DATA IN A PLUME DATA BANK

The main disadvantage of the plume data bank method is the size of the data bank whenever it has to be complete.

This size depends on the different parameters that govern the plume structure. Taking them all means that the total amount of data is.

$$A_t\ A_u\ A_e\ A_h\ A_{gh}\ A_q\ A_d \times 3600$$

Here the diameter d and the heat flow Q have been considered. In the Stümke formula [7] the diameter as well as the exit velocity are used with the temperature of the effluent. These data are related to the heat flow which is used in the formula proposed by Briggs in his basic paper [8].

Taking only 5 values per parameter leed to 3600×5^7 data, which is enormous. However, we could select the plumes necessary for a specific problem. This significantly reduces the memory requirements and eventually makes computation possible in a medium size computer.

ADVANTAGES OF THE USE OF A PLUME DATA BANK

The main advantage of the plume data bank are:

1. No data are stored concerning the coordinated and the intensity of the source. This mean that direct use of such a data bank permits computations for time-varying pollution sources. Variation may be in intensity as well as in space. This property has to be considered for the calculation in the case of plants that change over the special fuel at critical moments.
2. No data are stored concerning the wind direction. This is a less important advantage. Theoretically it is possible to calculate a chart corresponding to a weather condition at a well-determined moment with the correct angle. Practically, one has seldom to calculate particular situations with angles that are different from a multiple of 10 degrees, mostly because the measuring devices do not permit a higher accuracy.

ADVANTAGES OF THE USE OF POLLUTION CHARTS

The main advantages of the use of pollution charts are:

1. It seems more practical to use a databank of pollution charts because

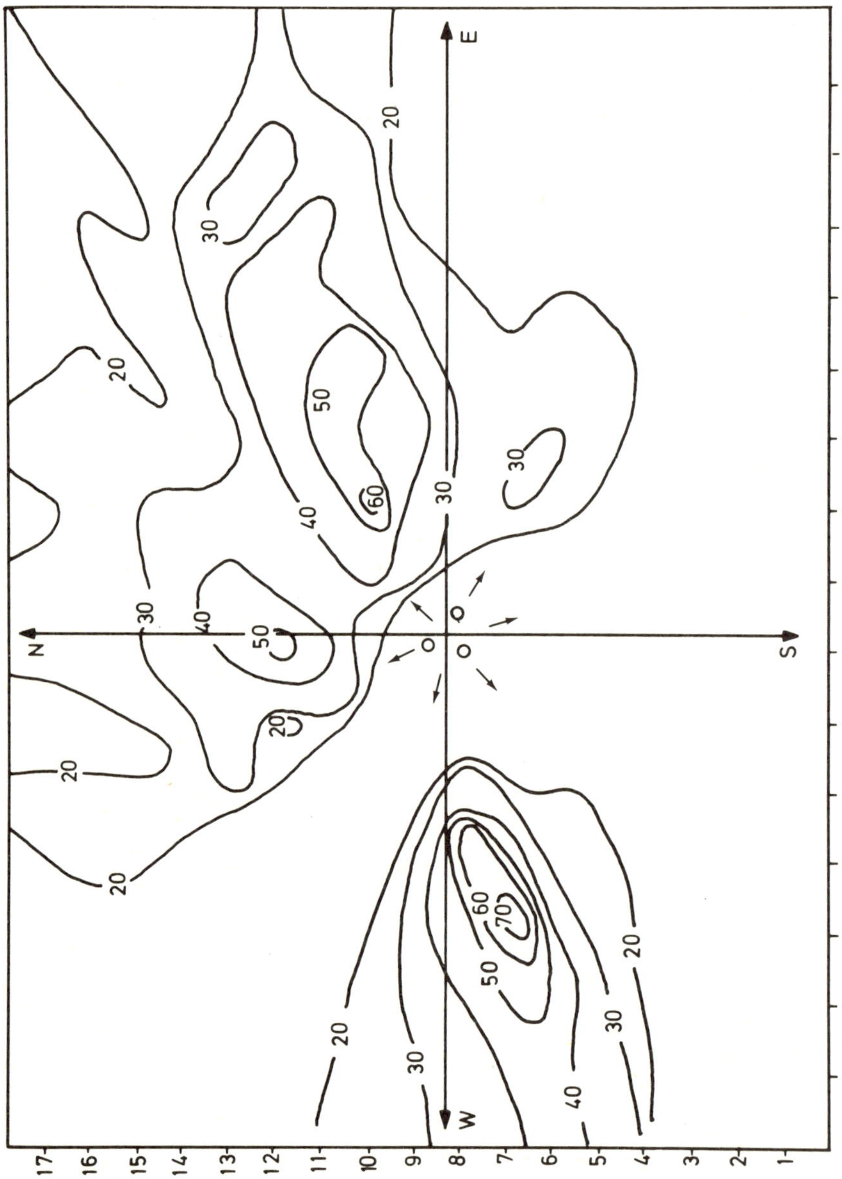

Fig. 13 - Result of code IFDM (95% values, 1 year, 24 hourly mean values)

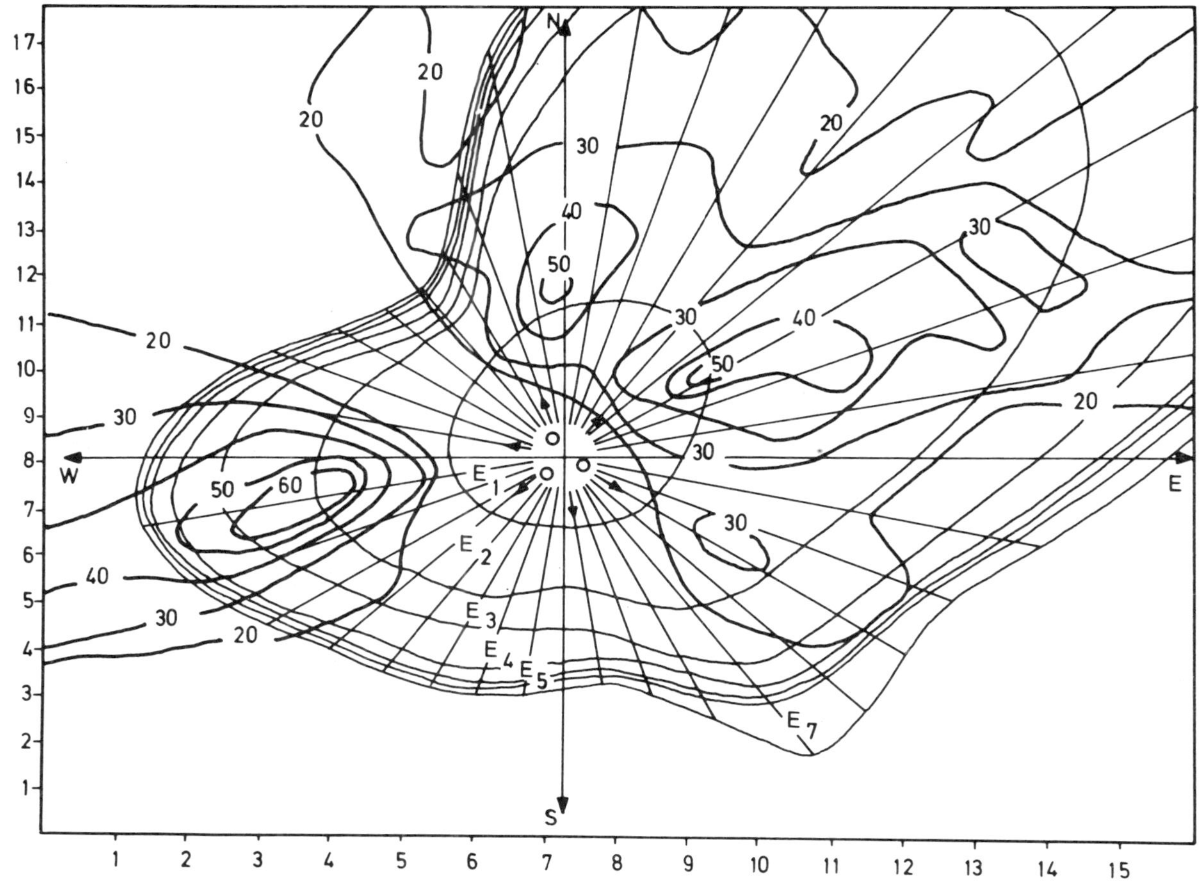

Fig. 14 - Superposition of pollution 95% distribution and wind direction distribution

such a data bank has to be set up for other reasons too e.g. the inspection of the pollution corresponding to a well-determined set of weather parameters. Such inspection is sometimes necessary to discover where the highest pollution level in a region is.

2. It is sometimes the only possible approach e.g: The calculation of distribution functions for ± 200 stacks in a region of 17 x 22 km starting from a plume data bank is practically impossible for two reasons.
   a. The supplementary inner loop (Fig. 12) changes the machine time from minutes to hours.
   b. The size of the disc memory has to be big enough to be able to contain all plume data. This reason, however, is not so limiting since one theoretically may select all necessary plumes as mentioned earlier.

CONCLUSIONS

Here two mean approaches are considered: the plume data bank and the precalculation of pollution charts.

- The plume data bank is a possible method for any computations where one or both loops in Fig. 12 is used only a few numbers of time. This means when the number of stacks is low or when averages or distributions are to be taken from a small set of meteorological situations.
- The plume data bank is practically unavoidable when the source is moving in time.
- Precalculation of pollution charts is often a necessity because it eliminates a loop in the flow chart (Fig. 12 with broken lines). When the number of stacks with time-dependent emission is limited the pollution charts may updated. This however, is possible a few times only, because updating is very time consuming when one does not have a plume data bank available. A disadvantage of the pollution chart method is that this data bank is region-dependent and one has to make a set of charts corresponding to each region under consideration. The number of charts depends on the discretisation. A reduction is possible when the Gaussian distribution is averaged over the angle of discretisation. However, the meteorological data bank must be contain data for the corresponding directions of the wind. This means that a measuring device giving wind directions every 20 degrees cannot be coupled to a program where the distribution has been averaged over 30 degrees. We have used, till now, the pollution chart approach because no moving sources have been considered and in

general we had to make pollution charts anyway for the preliminary studies.

Table IV summarises the advantages and disadvantages of both methods as a function of the problems.

The following code has been used:

TC = Time consuming

MC = Memory consuming

P = Possible

I = Impossible

Table IV

Advantages and disadvantages of the methods

problem	plume data bank	charts first	comments
Comparison of weather (fixed stack)	P P	P P	- -
1 chart (200 fixed stacks)	P,MC	P,TC	time is small
Yearly mean values (200 fixed stacks)	P,TC	P	-
Frequency distribution (200 fixed stacks)	TC	P	very time consuming
Moving sources	P	I	-

ACKNOWLEDGEMENTS

Interesting discussion have taken place with Mr.De Maere, Mr.Legrand Mr.Bhargava (India) and all members of the group M.M.

The calculations of Tables I, II and III as the Fig. 8, 9, 10 and 11 were made on the minicomputer of Mr.Beets (Safeguards)

REFERENCES

1. O.SUTTON, *Proc.Roy.Soc.(London) Ser. A135:143*

2. D.SLADE, "Meteorology and atomic energy 1968 U.S. Atomic Energy Commission

Technical Information", *Publication TID-214190*

3. H.BULTYNCK and L.MALET, *Tellus* 14 (1972), 5
4. LEGRAND et al, "Voorbereidende Studies, Automatische luchtveronreinigingsnetten voor het Antwerps havengebied en de Gentse Kanaalzone", *Rapport S.C.K.,* maart 1972
5. H.BULTYNCK, L.MALET, L.SHARMA, J.VAN DER PARREN, "Atmospheric dilution factors and calculation of doses in the environment of S.C.K./C.E.N. Mol for short and long stack discharges BLG 446, S.C.K./C.E.N., June 1970
6. G.BRIGGS, *Nuclear Safety, 12, No.1,* Jan.Feb., 1971
7. H.STUMKE, "Vorschlag einer empirische Formel für Schornsteinüberhöhung in Auschuss und Ueberprüfung bekannter Formeln mit zusätzlichem Beobachtüngs-Material", *Institut für Gasströmungen der technische Hochschule* - Stuttgart 1963
8. G.BRIGGS, "Plume rise", *USAEC no. TID-25075.*

USING Z-TRANSFORMS TO DETERMINE DIGITAL FILTERS FOR THE ON-LINE CALCULATION OF THE MEAN OF METEOROLOGICAL AND POLLUTION DATA

G.COCQUYT

SCK/CEN, Department Reactor Studies, Section Electronic Computers, Mol, Belgium

INTRODUCTION: TIME DEPENDENT SIGNALS, WHAT DO THEY LOOK LIKE, HOW ARE THEY DESCRIBED?

SIGNALS, WHAT THEY LOOK LIKE

In general two distinct types of signals are detected in air pollution.

A "smooth" signal, such as temperature

This signal looks like:

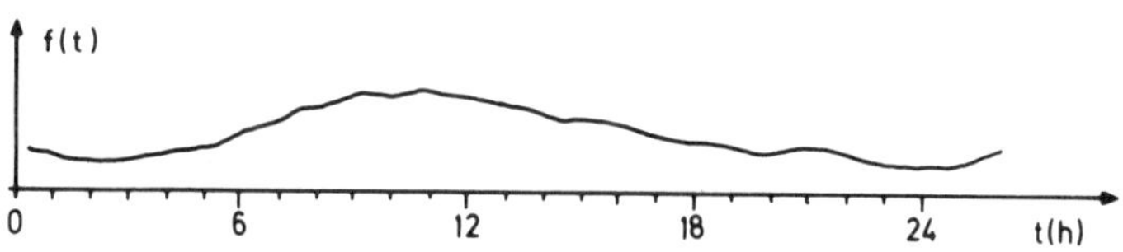

Fig. 1 - Smooth signal

A "noisy" signal such as windspeed

This signal looks like:

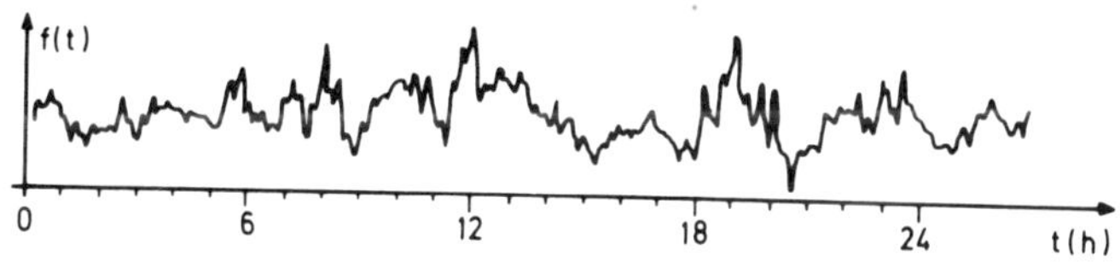

Fig. 2 - Noisy signal

Both signals are supposed to represent "reality". However, one may have noise corrupted signals. Sometimes the noise is of a different nature and has a higher frequency spectrum, while the signal itself is rather smooth.

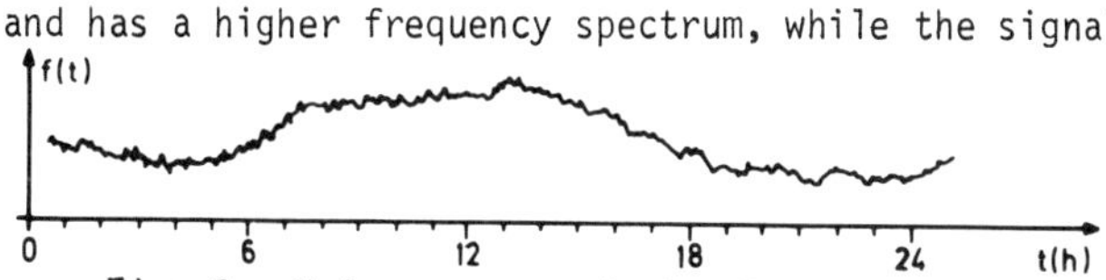

Fig. 3 - Noise corrupted signal

In this case the amplitude of the noise is smaller than the amplitude of the signal. The opposite is also possible (but difficult to drew).

In both these last cases it is simple to increase the signal-to-noise ratio. The problem is much more difficult when the frequency spectrum of the corrupting noise lies within the frequency spectrum of the signal to be detected.

SIGNALS, HOW CAN WE DESCRIBE THEM?

In general, two ways for the description of stochastic signals are used.

a. the auto-correlation function

b. the power spectrum.

a. The most simple way for the description of a stochastic signal is the auto-correlation function $\emptyset$ of that signal. The auto-correlation function gives, by definition, the correlation between two points of the same signal, seperated a time τ from each other.

A noisy signal normally contains higher frequencies. For such signals there is no correlation between two points, even when they are separated by a relatively short time from each other.

On the contrary, a smooth signal is a signal for which succeeding points are correlated to each other, even when the separation in time is rather high.

The auto-correlation function is given by:

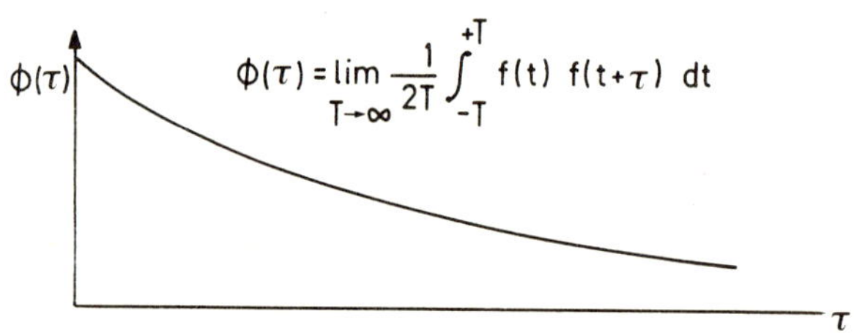

$$\phi(\tau) = \lim_{T \to \infty} \frac{1}{2T} \int_{-T}^{+T} f(t)\, f(t+\tau)\, dt$$

Fig. 4 - Auto-correlation function of a smooth signal

The auto-correlation function is (for non-periodic functions) maximum for τ= o, indeed each point is correlated the strongest with itself. This maximum is the mean squared value of the signal.

A steeply descending auto-correlation function indicates a noisy signal. A smooth signal corresponds to a slowly descending auto-correlation function. The three typical stochastic signals above are described by three typical auto-correlation functions (Fig.4,5,6)

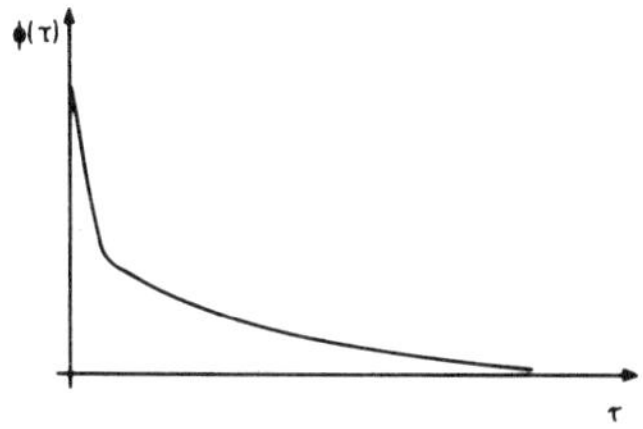

Fig. 5 - Auto-correlation of a noisy signal

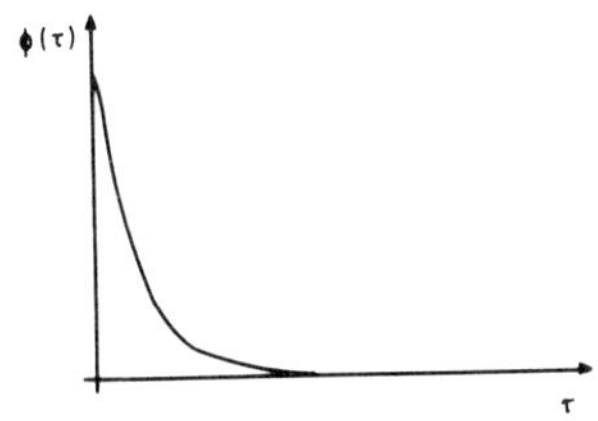

Fig. 6 - Auto-correlation of noise corrupted signal

b. The power spectrum. Another, more complicated way of describing signals is the use of power spectra, Power spectra give the "power" of a component of the signal is a well determined frequency band (infinitely small) as a function of the frequency. The total power is the integral of the power spectrum and is equal to the mean squared signal. A noisy signal has a rather large power spectrum, a smooth signal has a rather small spectrum (no high frequencies), a noise-corrupted signal may have a spectrum with two maxima, one for the signal and one for the noise.

The three typical stochastic signals above are described by three typical power spectra (Fig.7,8,9).

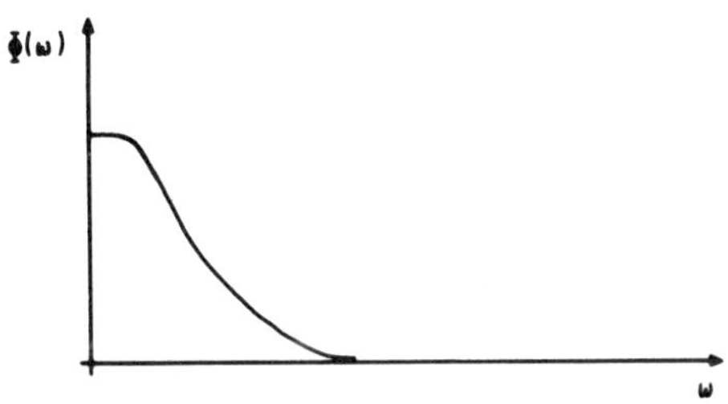

Fig. 7 - Power spectrum of a smooth signal

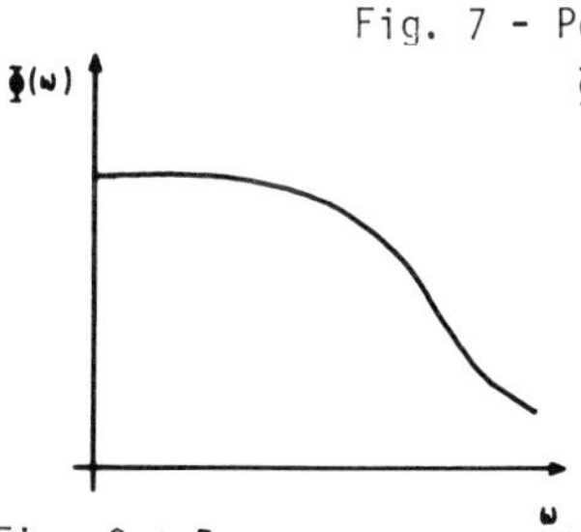

Fig. 8 - Power spectrum of a noisy signal

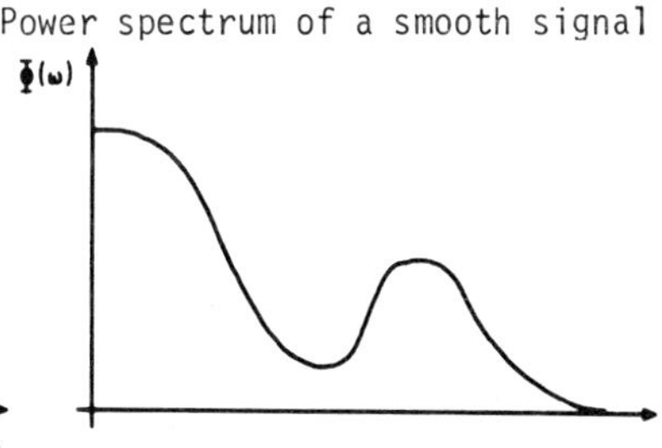

Fig. 9 - Power spectrum of a noise corrupted signal

c. Relation between the auto-correlation function and the power spectrum.

Both ways of describing stochastic signals are equivalent. It is possible to calculate the power spectrum from the auto-correlation function and vice-versa.

We have

$$\Phi(\omega) = \int_{-\infty}^{+\infty} \phi(\tau) e^{-j\omega\tau} d\tau$$

The power spectrum $\Phi(\omega)$ is the Fourier transform of the auto-correlation function $\phi(\tau)$. Using the inverse Fourier transform we can find the auto-correlation function from the power spectrum.

ON THE ON-LINE CALCULATION OF THE MEAN AND FILTERING

In general, data are "filtered" by calculating their mean value using for example the last 12 samples. This may be an hourly mean of data collected every five minutes.

Two questions arise:

1. What does this computation really mean?
2. Is this the best way of filtering, or, are the data to be filtered or not?

THE MEANING OF THE CALCULATION OF THE MEAN

Let us first consider a smooth signal. Such a signal has a time derivative which does not suddenly change. A standard curve with a constant time derivative is a straight line (Fig. 10).

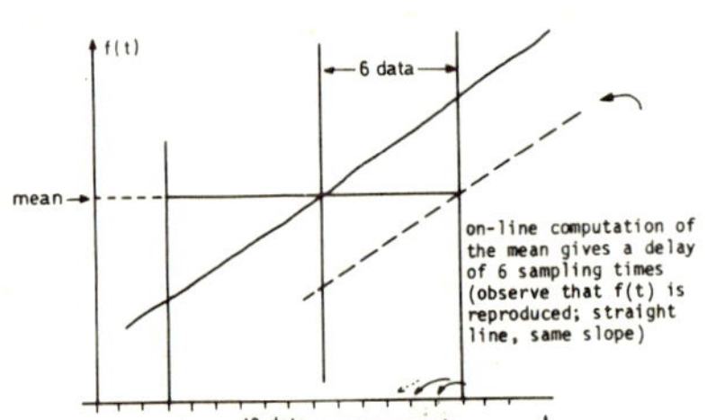

Fig. 10 - Response of a filter simulating the moving average

From the figure we can conclude that for f(t) being a straight line, the on-line computation of the mean of the N last values, leads to a reproduction of f(t) delayed by $\frac{N}{2}$ sampling times (on the figure N = 12).

Suppose now that f(t) is noisy and with a constant mean value (curve b). It is much more complicated to draw the result of the calculation of the mean, but it is easy to see that each new value is put together with the N-1 values collected earlier, and that each of these data are taken into account with a weighting factor $\frac{I}{N}$. In fact the computation of the mean in this case causes a decrease of the fluctuations by a factor N.

A curve f(t) which is composed of a low frequency component and a high frequency component, handled like this in general gives a decrease of the fluctuations by N and a delay for the mean value by $\frac{N}{2}$.

IS THIS THE BEST WAY OF FILTERING?

In order to answer this question one must ask another question: what criterion must be used? Sometimes this question can be answered easily. For example in the case of a noise corrupted signal, the criterion is that the signal to noise ratio must be as high as possible. Or, more exactly, the error e(t) between the filter output and the desired output must be minimal, for example (this is mostly the case) in the mean square sense.

We must have

$$\int_{o}^{\infty} e^2(t)\, dt \text{ minimal}$$

However, it is often very difficult to know exactly what the desired response is! Moreover, it is not always possible to define the signal! Maybe, for smooth signals, the best is not to filter at all.

We are not going to define more in detail what good filtering is, but we accept that the computation of the mean is the best way of filtering the data. Then there is no problem of defining the criterion.

THE EXACT SOLUTION OF THE PROBLEM

The criterion is that the filter must give the mean over the N last data as accurately as possible; the desired output signal of the filter is this mean f(t). In general when the mean over N data has to be calculated, one needs at least N memory places. The problem is to calculate this mean with a filter using a reasonably small amount of computer memory.

A block diagram indicates quite well what is wanted.

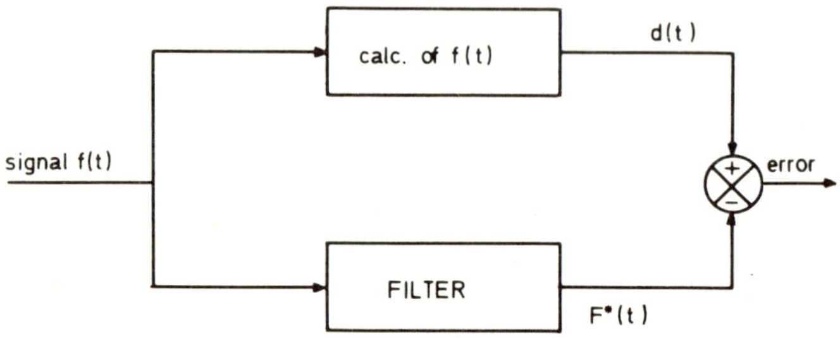

Fig. 11 - Flowchart for the determination of the optimal filter

THE SOLUTION IN THE CASE OF CONTINUOUS FILTERING

In fact this solution is not very important here because we intend to find reasonably good digital filters. However, often analog (continuous) signal conditioning modules are used before the entrance of the analog-to-digital converter (ADC). It is of interest to know how they disturb the signal and eventually to use this paper not only for the determination of digital filters but also find ideas of better filtering in an analog signal conditioning module (see Final Remark).

Simple RC filters

In general signal conditioning modules are used to increase the signal-to-noise ratio, or to decrease the frequency band of the power spectrum. Therefore mostly simple RC filters are used. It is important to know what such filters do. This can be described by their impulse response or differential equation.

We have:

diff. eq. : $T\frac{dV_o}{dt} + V_o = V_i(t)$

V_i = input signal V_o = output signal

Impulse response $h(t) = e^{-t/T}$

It is of interest to indicate that a straight line (rampfunction) for $V_i(t)$ is reproduced (after a transient of 4 T units of time) with a delay of T units of time, and that the spectrum is decreasing at a frequency equal to $\frac{1}{T}$. Moreover, in terms of resistors and capacitors T = RC.

A higher order filter corresponds to a higher order differential equation, we will consider here only higher order filters composed of a series of identical first order filters for simplicity.

Using the best continuous filters, the problem

Suppose that the filter that gives the exact mean ($\frac{1}{N}\int_0^N f(t)\, dt$) has an impulse response m(t), then the output d(t) (desired response) is given by

$$d(t) = \int_{-\infty}^{+\infty} m(t_1)\; f(t-t_1)\; dt_1$$

The output of the filters is

$$f^*(t) = \int_{-\infty}^{+\infty} h(t_1)\; f(t-t_1)\; dt_1$$

The error is

$$e(t) = d(t) - f^*(t) = \int_{-\infty}^{+\infty} (m(t_1) - h(t_1))\; f(t-t_1) dt_1$$

The integral of the square of this error must be minimized. This is possible, by calculating the partial derivatives of this integral, with respect to the filterparameters (T and number of filters eventually put in series) equal to zero and solving for this parameters.

The problem is that f(t) is a stochastic signal and, as such, not expressible as a function of time.

Theoretical solution

It is now possible to convert the equations mentioned above in terms of the auto-correlation function $\emptyset_{ff}(\tau)$ which describes the input signal of the filter.

Squaring the error we obtain

$$e^2 = d^2 - 2df^* + f^{*2}$$

Taking the limit $\bar{e}^2(t) = \lim_{T\to\infty} \frac{1}{2T} \int_{-T}^{+T} e^2(t)\, dt$ we have

$$\overline{e^2(t)} = \overline{d^2(t)} - 2 \lim_{T\to\infty} \frac{1}{2T} \int_{-T}^{+T} dt \int_{\infty}^{+\infty} dt_1\; h(t_1)\; f(t-t_1)\; d(t) \qquad +$$

$$+ \lim_{T\to\infty} \frac{1}{2T} \int_{-T}^{+T} dt \int_{-\infty}^{+\infty} dt_1\, h(t_1) f(t-t_1) \int_{-\infty}^{+\infty} dt_2 h(t_2) f(t-t_2)$$

Introducing the auto-correlation function we find

$$e^2(t) = \phi_{dd}(0) - 2 \int_{-\infty}^{+\infty} dt_1 h(t_1) \phi_{fd}(t_1) + \int_{-\infty}^{+\infty} dt_1 h(t_1) \int_{-\infty}^{+\infty} dt_2 h(t_2) \phi_{ff}(t_1 - t_2)$$

where $\phi_{fd}(t_1)$ is the cross-correlation of the input signal f(t) with the ideal mean calculator output giving $d(t) = \frac{1}{X} \int_0^X f(t)\, dt$, defined by

$$\phi_{fd}(t_1) = \lim_{T\to\infty} \frac{1}{2T} \int_{-T}^{+T} d(t) \,.\, f(t + t_1) dt$$

ϕ_{ff} is the auto-correlation function of the input signal f(t).

Taking the partial derivative with respect to the filter parameters T, and n (the order of the filter) (they can be found in $h(t_1)$) and putting them equal to zero leads to a set of equations which permit the determination of the optimal T and n, provided ϕ_{ff} is given.

It is possible to simplify the solution using Fourrier-transforms, then power spectra are introduced. The integral of the power spectrum is the mean squared value of the signal. We then have to minimize the following integral:

$$\int_0^{\infty} \left| \frac{1}{N} \; \frac{1 - e^{-j\omega N}}{j\omega} - \frac{1}{(1 + Tj\omega)n} \right|^2 \Phi_{ff}(\omega)\, d\omega$$

with $\Phi_f(\omega)$ = power spectrum of the input signal.

If the auto-correlation function ϕ_{ff} is known, the power spectrum Φ_{ff} is known too. Once this given, the problem can be solved exactly. However, it is reasonable to draw not too much attention, to the exact solution which depends on Φ_{ff} (ω). Moreover the computations are tedious.

Here we will try to design "the best filter", in a completely different way. We will do this by heuristic thinking on types of signals and on what the filter has to do. Whenever possible, we will try to get a filter of good preformance independent of the auto-correlation function of the input signal.

SAMPLED DATA SYSTEMS

The method above, valid for continuous signals, may be adapted for discrete signals or sampled data. One simply has to replace (in principle) the integrals by summations. However, it is possible to do it in a simpler way.

Z-TRANSFORMS

Definitions

Let us consider a time series of collected (discrete) data

$$f(0),\ f(1),\ f(2)\ .\ .\ .\ f(k)\ .\ .\ .\ f(x-1)$$

and define (for simplicity) the corresponding Z-transform as:

$$F(Z) = f(0)Z^{0} + f(1)Z^{-1} + f(2)Z^{-2} + \ .\ .\ .\ f(k)Z^{-k} + \ .\ .\ .\ f(x-1)Z^{-x+1}$$

The multiplication factor Z^{-k} indicates the time, and must be considered as an operator performing a shift of k units back in time.

Transfer functions in the Z-transform domain and recurrence relation

The Z-transform has many properties close to the Laplace transform [1,2] and the Z-variable plays more or less the same role as the Laplace variable s. So a transfer function for a digital filter may be defined as the ratio of the Z-transform of the output to the Z-transform of the input. Moreover, this ratio is the Z-transform of the impuls response of the filter. So an integrator (with impulse response $1 + Z^{-1} + Z^{2} + ...$) has a transfer function $\frac{1}{1 - Z^{-1}}$.

In general transformed functions are indicated by capital letters (F(Z) is the transform of f(t)).

The sampled data transfer function is related to simple digital computer programming algorithms.

So a transfer function

$$\frac{F^{*}(Z)}{F\ (Z)} = \frac{\beta + \beta_1 Z^{-1} + \beta_2 Z^{-2} + \ldots}{1 - \alpha_1 Z^{-1} - \alpha_2 Z^{-2}\ \ldots.} = \frac{\sum_{j=0}^{n} \beta_j Z^{-j}}{1 - \sum_{i=1}^{m} \alpha_i Z^{-i}}$$

corresponds to the simple relation:

$$f^{*}(k) = \sum_{i=1}^{m} \alpha_i\ f^{*}(k-i) + \sum_{j=0}^{n} \beta_j f(k-j)$$

This relation has been discussed in ref. 3 .

This recurrence equation will be used for the determination of simple almost memoryless filters for the calculation of the mean.

RESPONSES OF DIGITAL FILTERS

DIGITAL FILTERS CALCULATING THE MEAN OVER N DATA

The output of such filters is found to be

$$\overline{f(k+1)} = \overline{f(k)} - \frac{1}{N} f(k+1-N) + \frac{1}{N} f(k+1)$$

This means:

1. the last data, $f(k+1)$, is taken into account with a factor $\frac{1}{N}$;
2. the data collected N sampling times before, $f(k+1-N)$, used for the calculation of the previous mean, is dropped:
3. all the data in between; gathered in $f(k)$, remain.

Using the relation to transfer functions we find that the transfer function of such a filter is

$$\frac{\overline{F(Z)}}{f(Z)} = \frac{1}{N} \frac{(1 - Z^{-N})}{(1 - Z^{-1})}$$

The response of this filter has been analysed before.

The mean properties are:

1. a devision of the new coming data by N
2. a shift in time for a smooth signal by $\frac{N}{2}$.

More details about the response of this filter are given in ref. 4.

SIMPLE DIGITAL FILTERS

The most simplest digital filters giving an unbiased "mean" has an algorithm

$$f^* (K) = (1 - \frac{1}{N})\, f^* (k-1) + \frac{1}{N} f(k)$$

Here the same symbol N has been used in order to draw attention to the fact that such a filter may divide new coming data by the same factor (N), just like a filter calculating the mean.

This filter which considers the previous result, $f^*(k-1)$, as equivalent to N-1 data is adding the new coming data, f(k), and calculating the mean.

Such a digital filter has a transfer function

$$\frac{F^*(Z)}{F(Z)} + \frac{1/N}{1-(1-1/N)Z^{-1}}$$

The impulse response is an exponentially decaying function, very similar to that of the continuous RC filter as discussed before. Therefore the time-constant RC must be such that $e^{-t/RC} = 1 - \frac{1}{N}$ for t = sampling time. Continuous RC filters follow a rampfunction with a delay of RC seconds. Digital filters, having more or less [5] the same properties have a delay of N sampling times.

ALMOST MEMORYLESS DIGITAL FILTERS FOR THE CALCULATION OF THE MEAN

THE PROBLEM

The problem is to find a simple filter using practically no memory and calculating the means "as close as possbile". Knowing that the exact solution is only possible for signals with well known auto-correlation functions, the problem is to find a simple digital filter that may work nearly optimally even with signals described by different auto-correlation functions.

A POSSIBLE FILTER FOR SMOOTH SIGNALS

As mentioned before a first order filter (only Z^{-1} and not higher orders of it in the denominator of the transferfunction) may be sufficient for smooth signals when the delay is well adjusted.

Indeed for smooth signals the delay is the most important characteristic of the calculation of the mean. This means, when the calculation is to be performed over N values, the first order digital filter must be:

$$\frac{F^*(Z)}{F(Z)} = \frac{1/(N/2)}{1-(1-1/(N/2))Z^{-1}}$$

Such a filter has the same delay as the calculation of the mean (N/2). Its algorithm is:

$$f^*(k) = (1 - 1/N/2)) \quad f^*(k-1) + \frac{1}{(N/2)} f(k)$$

A POSSIBLE FILTER FOR NOISY SIGNALS

The calculation of the mean, for such signals, essentially involves a division of the fluctuations around the mean by the number of data taken into account. It is logical to use a filter that performs the same division. This means a first order filter which the algorithm is

$$f^* (k) = (1 - \frac{1}{N}) f^* (k-1) + \frac{1}{N} f(k)$$

A "PASSE-PARTOUT" FILTER

Such a filter must have a delay (for the smooth part of f(k)) of $\frac{N}{2}$ and an attenuation of N (for the noisy part).

It is clear that a first order filter of the type described above cannot do the job.

However, a second order filter composed of two identical filters put in series may give better results. In general, a filter of that type has a transfer function.

$$\frac{F^* (Z)}{F (Z)} = \left(\frac{1/A}{1 - (1-1/A)Z^{-1}}\right)^n$$

where A is to be adapted for either correct shift in time or correct diminuation of the fluctuations.

For a correct shift in time $(\frac{N}{2})$ we have:

$$\frac{F^* (Z)}{F (Z)} = \left(\frac{1/(N/2n)}{1 - (1-1/N/2n)Z^{-1}}\right)^n$$

Indeed the shift of the first order filter is given by N/2n units of time (sampling instants) and there are n first order filters put in series.

For a correct gain (1/N) we have:

$$\frac{F^* (Z)}{F (Z)} = \left(\frac{\sqrt[n]{1/N}}{1 - (1-\sqrt[n]{1/N})Z^{-1}}\right)^n$$

Indeed each input data is divided for the first order filter by $\sqrt[n]{N}$, and there are n filters put in series.

The "passe partout" filter must have the best possible delay together with the best possible gain. This depends on the number of values N.

A program has been made for the computation of a table giving the gain for filters with a correct delay (Table I) and an other for the computation of the table giving the delay of filters with correct gain (Table II).

A third program has been made plotting graphically the gain (for the first) as well as the delay (for the second) as a function of the order of the filter.

Three outputs are represented here (for N = 12, 20 and 25).

Looking at the result for 12 measurements, we find that the delay is maximum ($\pm$ 7 sampling times) for a second order filter, and that with such a filter the gain as well as the delay (ideal: 6 sampling times) is near to the optimal values. Such a filter must be able to do the job for an important variety of auto-correlation functions.

Table I - Attenuation of filter with given delay for ramp input (/1ØØØ)

ORDER OF FILTER	1	2	3	4	5	6
GAIN FOR N= 5	= 0.003	= 0.002	= 0.001	= 0.000	= 0.000	= 0.000
GAIN FOR N= 10	= 0.005	= 0.006	= 0.005	= 0.003	= 0.001	= 0.000
GAIN FOR N= 15	= 0.008	= 0.014	= 0.016	= 0.012	= 0.008	= 0.004
GAIN FOR N= 20	= 0.010	= 0.025	= 0.037	= 0.039	= 0.032	= 0.022
GAIN FOR N= 25	= 0.013	= 0.039	= 0.072	= 0.095	= 0.098	= 0.082
GAIN FOR N= 30	= 0.015	= 0.056	= 0.125	= 0.198	= 0.243	= 0.244
GAIN FOR N= 35	= 0.018	= 0.077	= 0.199	= 0.366	= 0.525	= 0.616
GAIN FOR N= 40	= 0.020	= 0.100	= 0.296	= 0.625	= 1.024	= 1.372
GAIN FOR N= 45	= 0.023	= 0.127	= 0.422	= 1.001	= 1.845	= 2.781
GAIN FOR N= 50	= 0.025	= 0.156	= 0.579	= 1.526	= 3.125	= 5.233
GAIN FOR N= 55	= 0.028	= 0.189	= 0.770	= 2.234	= 5.033	= 9.270
GAIN FOR N= 60	= 0.030	= 0.225	= 1.000	= 3.164	= 7.776	= 15.63
GAIN FOR N= 65	= 0.033	= 0.264	= 1.272	= 4.358	= 11.60	= 25.26
GAIN FOR N= 70	= 0.035	= 0.306	= 1.588	= 5.862	= 16.81	= 39.40
GAIN FOR N= 75	= 0.038	= 0.352	= 1.953	= 7.725	= 23.73	= 59.61
GAIN FOR N= 80	= 0.040	= 0.400	= 2.370	= 10.00	= 32.77	= 87.79
GAIN FOR N= 85	= 0.043	= 0.452	= 2.843	= 12.75	= 44.37	= 126.3
GAIN FOR N= 90	= 0.045	= 0.506	= 3.375	= 16.02	= 59.05	= 178.0
GAIN FOR N= 95	= 0.048	= 0.564	= 3.969	= 19.89	= 77.38	= 246.2
GAIN FOR N= 100	= 0.050	= 0.625	= 4.630	= 24.41	= 100.0	= 334.9
GAIN FOR N= 105	= 0.053	= 0.689	= 5.359	= 29.68	= 127.6	= 448.8
GAIN FOR N= 110	= 0.055	= 0.756	= 6.162	= 35.75	= 161.1	= 593.3
GAIN FOR N= 115	= 0.058	= 0.827	= 7.041	= 42.70	= 201.1	= 774.6
GAIN FOR N= 120	= 0.060	= 0.900	= 8.000	= 50.63	= 248.8	= 1000
GAIN FOR N= 125	= 0.063	= 0.977	= 9.042	= 59.61	= 305.2	= 1278
GAIN FOR N= 130	= 0.065	= 1.056	= 10.17	= 69.73	= 371.3	= 1617
GAIN FOR N= 135	= 0.068	= 1.139	= 11.39	= 81.09	= 448.4	= 2027
GAIN FOR N= 140	= 0.070	= 1.225	= 12.70	= 93.79	= 537.8	= 2522
GAIN FOR N= 145	= 0.073	= 1.314	= 14.11	= 107.9	= 641.0	= 3113
GAIN FOR N= 150	= 0.075	= 1.406	= 15.63	= 123.6	= 759.4	= 3815

*

Table II - Delay of filter with given attenuation for noisy input

ORDER OF FILTER	1	2	3	4	5	6
DELAY FOR N= 5	= 5.000	= 4.472	= 5.130	= 5.981	= 6.899	= 7.846
DELAY FOR N= 10	= 10.00	= 6.325	= 6.463	= 7.113	= 7.925	= 8.807
DELAY FOR N= 15	= 15.00	= 7.746	= 7.399	= 7.872	= 8.594	= 9.423
DELAY FOR N= 20	= 20.00	= 8.944	= 8.143	= 8.459	= 9.103	= 9.885
DELAY FOR N= 25	= 25.00	= 10.00	= 8.772	= 8.944	= 9.518	= 10.26
DELAY FOR N= 30	= 30.00	= 10.96	= 9.322	= 9.361	= 9.872	= 10.58
DELAY FOR N= 35	= 35.00	= 11.83	= 9.813	= 9.729	= 10.18	= 10.85
DELAY FOR N= 40	= 40.00	= 12.65	= 10.26	= 10.06	= 10.46	= 11.10
DELAY FOR N= 45	= 45.00	= 13.42	= 10.67	= 10.36	= 10.71	= 11.32
DELAY FOR N= 50	= 50.00	= 14.14	= 11.05	= 10.64	= 10.93	= 11.52
DELAY FOR N= 55	= 55.00	= 14.83	= 11.41	= 10.89	= 11.15	= 11.70
DELAY FOR N= 60	= 60.00	= 15.49	= 11.75	= 11.13	= 11.34	= 11.87
DELAY FOR N= 65	= 65.00	= 16.13	= 12.06	= 11.36	= 11.52	= 12.03
DELAY FOR N= 70	= 70.00	= 16.73	= 12.36	= 11.57	= 11.70	= 12.18
DELAY FOR N= 75	= 75.00	= 17.32	= 12.65	= 11.77	= 11.86	= 12.32
DELAY FOR N= 80	= 80.00	= 17.89	= 12.93	= 11.96	= 12.01	= 12.46
DELAY FOR N= 85	= 85.00	= 18.44	= 13.19	= 12.15	= 12.16	= 12.58
DELAY FOR N= 90	= 90.00	= 18.97	= 13.45	= 12.32	= 12.30	= 12.70
DELAY FOR N= 95	= 95.00	= 19.49	= 13.69	= 12.49	= 12.43	= 12.82
DELAY FOR N= 100	= 100.0	= 20.00	= 13.93	= 12.65	= 12.56	= 12.93
DELAY FOR N= 105	= 105.0	= 20.49	= 14.15	= 12.81	= 12.68	= 13.03
DELAY FOR N= 110	= 110.0	= 20.98	= 14.38	= 12.96	= 12.80	= 13.13
DELAY FOR N= 115	= 115.0	= 21.45	= 14.59	= 13.10	= 12.92	= 13.23
DELAY FOR N= 120	= 120.0	= 21.91	= 14.80	= 13.24	= 13.03	= 13.33
DELAY FOR N= 125	= 125.0	= 22.36	= 15.00	= 13.38	= 13.13	= 13.42
DELAY FOR N= 130	= 130.0	= 22.80	= 15.20	= 13.51	= 13.24	= 13.51
DELAY FOR N= 135	= 135.0	= 23.24	= 15.39	= 13.64	= 13.34	= 13.59
DELAY FOR N= 140	= 140.0	= 23.67	= 15.58	= 13.76	= 13.43	= 13.67
DELAY FOR N= 145	= 145.0	= 24.08	= 15.76	= 13.88	= 13.53	= 13.75
DELAY FOR N= 150	= 150.0	= 24.50	= 15.94	= 14.00	= 13.62	= 13.83
DELAY FOR N= 155	= 155.0	= 24.90	= 16.12	= 14.11	= 13.71	= 13.91
DELAY FOR N= 160	= 160.0	= 25.30	= 16.29	= 14.23	= 13.80	= 13.98
DELAY FOR N= 165	= 165.0	= 25.69	= 16.46	= 14.34	= 13.88	= 14.05
DELAY FOR N= 170	= 170.0	= 26.08	= 16.62	= 14.44	= 13.97	= 14.12
DELAY FOR N= 175	= 175.0	= 26.46	= 16.78	= 14.55	= 14.05	= 14.19
DELAY FOR N= 180	= 180.0	= 26.83	= 16.94	= 14.65	= 14.13	= 14.26
DELAY FOR N= 185	= 185.0	= 27.20	= 17.10	= 14.75	= 14.20	= 14.32
DELAY FOR N= 190	= 190.0	= 27.57	= 17.25	= 14.85	= 14.28	= 14.39
DELAY FOR N= 195	= 195.0	= 27.93	= 17.40	= 14.95	= 14.36	= 14.45
DELAY FOR N= 200	= 200.0	= 28.29	= 17.55	= 15.04	= 14.43	= 14.51

Looking in the same figure (graphical plotting 1) a maximum is found for the attenuation of those filters which have a correct delay ($(N/2n)^n$ has been plotted).

This attenuation is 9 (the desired value is 12).

Anyhow, both filters are capable of giving good results. Where the attenuation is more important such as for wind speed meters one can use:

$$\frac{F^*(Z)}{F(Z)} = \frac{1/N}{(1-(1-\sqrt{1/N})Z^{-1})^2}$$

For smooth signals such as for noise free temperature meter outputs one can use

$$\frac{F^*(Z)}{F(Z)} = \left(\frac{4/N}{1-(1-4/N)Z^{-1}}\right)^2$$

Both however, must give more or less equivalent results.

CHECKING THE FILTERS

A program in Fortran has been made by Bosmans and Van De Craen simulating such filters working on temperature data received from the meteorological service as the S.C.K.

Three filters with correct delay have been checked giving an error with respect to the mean. A typical result for a smooth signal is (App.1):

First order, Transfer function $\frac{1/6}{1-5/6\,Z^{-1}}$ error 0.186°C

Second order, Transfer function $\left(\frac{1/3}{1-2/3\,Z^{-1}}\right)^2$ error 0.143°C

Third order, Transfer function $\left(\frac{1/2}{1-1/2\,Z^{-1}}\right)^3$ error 0.227°C

That the third order filter gives a result that is worse than a second order filter, was foreseen. Indeed graphical plotting 1 gives attenuation factors that are not so good.

Three filters with correct attenuation have been checked on the same set of data. The results were (Appendix 2):

First order, Transfer function $\dfrac{1/12}{1 - 11/12\ Z^{-1}}$ error 0.389°C

Second order, Transfer function $\dfrac{0.09}{1-1.4Z^{-1}+.49Z^{-2}}$ error 0.137°C

Third order, Transfer function $\left(\dfrac{1/2.29}{1 - 1.29Z^{-1}}\right)^3$ error 0.170°C

FINAL REMARKS

It is now possible to go back to continuous filtering and remark that even this analysis in the Z-transform domain (for example indicating that a new coming data is divided by a certain value, which is difficult to detect in the continuous time world) may lead to a continuous time filter (double RC filter) whose output must be, for a large range of auto-correlation functions, very near to that of an integrator.

It is not difficult to extend this method so as to get an algorithm that gives the correct delay as well as the correct gain. However, the increase in precision in the windspeed and temperature data, is negligible.

ACKNOWLEDGEMENTS

The importance of the use of Z-transforms was indicated by Mr.Beets and made possible the tackling of the problem of on-line calculation of mean values in this way.

Encouraging discussions have taken place with Mr.De Maere, Mr.Kretzschmar, Mr.De Greef and the members of the group P.M.

REFERENCES

1. E.I.JURY, Sampled Data Control Systems, *J.Wiley*, 1958
2. JA.Z.ZYPKIN, Theorie der linearen Impulssysteme, *Oldenburg Verlag, 1967*
3. G.COCQUYT, Recursive betrekkingen voor digitale filters uitgaande van Z-transforms, *Het Ingenieursblad* nr. 21, 1970
4. G.COCQUYT, Antwoorden van een digitaal filter voor voortschrijdende uitmiddeling, *Het Ingenieursblad* nr.3, 1971
5. G.COCQUYT, Antwoorden van een exponentieel vergetend digitaal filter, *Het Ingenieursblad* nr. 24, 1970

Graphical plotting 1 - Graphical plotting of a filter with a given attenuation and a given delay as a function of N

"DELAY" = Delay of filter with given attenuation for noisy input

"GAIN" = Attenuation of filter with given delay for ramp input

Number of measurements: 12

Delay: 6

Attenuation: 12

ORDER

```
N= 1.0   DELAY= 12.000                                         *
N= 1.0   GAIN= 6.0000---------------------+
N= 1.5   DELAY= 7.8622                           *
N= 1.5   GAIN= 8.0000---------------------------+
N= 2.0   DELAY= 6.9282                        *
N= 2.0   GAIN= 9.0000-------------------------------+
N= 2.5   DELAY= 6.7548                       *
N= 2.5   GAIN= 8.9234------------------------------+
N= 3.0   DELAY= 6.8683                       *
N= 3.0   GAIN= 8.0000---------------------------+
N= 3.5   DELAY= 7.1188                        *
N= 3.5   GAIN= 6.5962----------------------+
N= 4.0   DELAY= 7.4449                         *
N= 4.0   GAIN= 5.0625-----------------+
N= 4.5   DELAY= 7.8168                           *
N= 4.5   GAIN= 3.6494-------------+
N= 5.0   DELAY= 8.2188                            *
N= 5.0   GAIN= 2.4883---------+
N= 5.5   DELAY= 8.6413                             *
N= 5.5   GAIN= 1.6138------+
N= 6.0   DELAY= 9.0785                               *
N= 6.0   GAIN= 1.0000----+
N= 6.5   DELAY= 9.5267                                *
N= 6.5   GAIN= 0.5944--+
N= 7.0   DELAY= 9.9831                                  *
N= 7.0   GAIN= 0.3399--+
N= 7.5   DELAY= 10.446                                   *
N= 7.5   GAIN= 0.1876-+
N= 8.0   DELAY= 10.914                                     *
N= 8.0   GAIN= 0.1001-+
N= 8.5   DELAY= 11.386                                      *
N= 8.5   GAIN= 0.0518-+
N= 9.0   DELAY= 11.862                                        *
N= 9.0   GAIN= 0.0260-+
N= 9.5   DELAY= 12.340                                          *
N= 9.5   GAIN= 0.0127-+
N= 10   DELAY= 12.821                                           *
N= 10   GAIN= 0.0061-+
N= 11   DELAY= 13.304                                             *
N= 11   GAIN= 0.0028-+*
```

Graphical plotting 2 - Graphical plotting of a filter with a given attenuation and a given delay as a function of N

"DELAY" = Delay of filter with given attenuation for noisy input

"GAIN" = Attenuation of filter with given delay for ramp input

Number of measurements: 20

Delay: 110

Attenuation: 20

ORDER

```
N= 1.0   DELAY= 20.000                                                  *
N= 1.0   GAIN= 10.000----------+
N= 1.5   DELAY= 11.052                                      *
N= 1.5   GAIN= 17.213------------------+
N= 2.0   DELAY= 8.9443                               *
N= 2.0   GAIN= 25.000-------------------------+
N= 2.5   DELAY= 8.2861                             *
N= 2.5   GAIN= 32.000--------------------------------+
N= 3.0   DELAY= 8.1433                             *
N= 3.0   GAIN= 37.037--------------------------------------+
N= 3.5   DELAY= 8.2374                             *
N= 3.5   GAIN= 39.424----------------------------------------+
N= 4.0   DELAY= 8.4590                              *
N= 4.0   GAIN= 39.063----------------------------------------+
N= 4.5   DELAY= 8.7565                               *
N= 4.5   GAIN= 36.353-------------------------------------+
N= 5.0   DELAY= 9.1028                                *
N= 5.0   GAIN= 32.000--------------------------------+
N= 5.5   DELAY= 9.4823                                 *
N= 5.5   GAIN= 26.792---------------------------+
N= 6.0   DELAY= 9.8853                                  *
N= 6.0   GAIN= 21.434----------------------+
N= 6.5   DELAY= 10.306                                    *
N= 6.5   GAIN= 16.446-----------------+
N= 7.0   DELAY= 10.739                                     *
N= 7.0   GAIN= 12.143-------------+
N= 7.5   DELAY= 11.182                                       *
N= 7.5   GAIN= 8.6505---------+
N= 8.0   DELAY= 11.634                                        *
N= 8.0   GAIN= 5.9605------+
N= 8.5   DELAY= 12.092                                          *
N= 8.5   GAIN= 3.9805----+
N= 9.0   DELAY= 12.555                                           *
N= 9.0   GAIN= 2.5812---+
N= 9.5   DELAY= 13.022                                             *
N= 9.5   GAIN= 1.6279--+
N= 10   DELAY= 13.493                                              *
N= 10   GAIN= 1.0000--+
N= 11   DELAY= 13.967                                                *
N= 11   GAIN= 0.5991-+*
```

Graphical plotting 3 - Graphical plotting of a filter with a given attenuation and a given delay as a function of N

"DELAY" = Delay of filter with given attenuation for noisy input

"GAIN" = Attenuation of filter with given delay for ramp input

Number of measurements: 25

Delay: 12.5

Attenuation: 25

ORDER

```
N= 1.0   DELAY= 25.000                                          *
N= 1.0   GAIN= 12.500-------+
N= 1.5   DELAY= 12.825                      *
N= 1.5   GAIN= 24.056-------------+
N= 2.0   DELAY= 10.000                 *
N= 2.0   GAIN= 39.063--------------------+
N= 2.5   DELAY= 9.0598                *
N= 2.5   GAIN= 55.902----------------------------+
N= 3.0   DELAY= 8.7721               *
N= 3.0   GAIN= 72.338-------------------------------------+
N= 3.5   DELAY= 8.7797               *
N= 3.5   GAIN= 86.039--------------------------------------------+
N= 4.0   DELAY= 8.9443               *
N= 4.0   GAIN= 95.367------------------------------------------------+
N= 4.5   DELAY= 9.2017                *
N= 4.5   GAIN= 99.229--------------------------------------------------+
N= 5.0   DELAY= 9.5183                *
N= 5.0   GAIN= 97.656-------------------------------------------------+
N= 5.5   DELAY= 9.8749                 *
N= 5.5   GAIN= 91.413----------------------------------------------+
N= 6.0   DELAY= 10.260                  *
N= 6.0   GAIN= 81.762-----------------------------------------+
N= 6.5   DELAY= 10.666                  *
N= 6.5   GAIN= 70.142------------------------------------+
N= 7.0   DELAY= 11.087                   *
N= 7.0   GAIN= 57.901-----------------------------+
N= 7.5   DELAY= 11.520                    *
N= 7.5   GAIN= 46.117------------------------+
N= 8.0   DELAY= 11.963                    *
N= 8.0   GAIN= 35.527------------------+
N= 8.5   DELAY= 12.413                     *
N= 8.5   GAIN= 26.526--------------+
N= 9.0   DELAY= 12.870                      *
N= 9.0   GAIN= 19.231----------+
N= 9.5   DELAY= 13.331                       *
N= 9.5   GAIN= 13.560-------+
N= 10   DELAY= 13.797                       *
N= 10   GAIN= 9.3132-----+
N= 11   DELAY= 14.267                        *
N= 11   GAIN= 6.2383----+*
```

APPENDIX 1

Program for the analysis of the filters with "correct delay"

XMEAN = Desired Value

V = Variance (for 12 measurements)

Z = Response first order filter

Var = Response first order filter for the estimation of the variance

Y2 = Response second order filter

Var2 = Response second order filter for the estimation of the variance

T3 = Response third order filter

Dev3 = Response third order filter for the estimation of the variance

```
C INITIALISATION
      REAL*8 TIME
      DIMENSION X(12)
      J=C
      Z=8.6
      VAR=0.17
      Y1=8.6
      Y2=8.6
      VAR1=0.17
      VAR2=0.17
      T1=8.6
      T2=8.6
      T3=8.6
      DEV1=0.17
      DEV2=0.17
      DEV3=0.17
      V1=0.
      V2=0.
      V3=0.
      D1=0.
      D2=0.
      D3=0.
      WRITE(6,15)
15    FORMAT(1H1,T1C,'TIME',T2C,'XMEAN',T3C,'V',T4C,'Z',T5C,
     +T6C,'Y2',T7C,'VAR2',T83,'T3',T90,'DEV3'///)
C INPUT CARDS
5     READ(5,1C)((X(K),K=1,12),TIME)
1C    FORMAT(12F5.2,1CX,A8)
C CALCULATION OF CLASSICAL MEAN AND VARIANCE PER HOUR
      XMEAN=C.
      V=0.
      DC 2C K=1,12
   2C XMEAN=XMEAN+X(K)
      XMEAN=XMEAN/12.
      DC 25 K=1,12
   25 V=V+(XMEAN-X(K))**2
      V=V/12.
C CALCULATICN (ALGORITHMES) FOR MEANS(Z AND Y2)
C AND VARIANCES(VAR AND VAR2) PER HOUR
      I=1
C EERSTE-ORDER FILTER  (FIRST ORDER FILTER)
  3C  Z=(Z*5.+X(I))/6.
      VAR=(VAR*5.+(X(I)-Z)**2)/6.
C TWEEDE-ORDER FILTER  (SECOND ORDER FILTER)
      Y1=(Y1*0.7)+(0.3*X(I))
      Y2=(Y2*0.7)+(C.3*Y1)
      VAR1=(VAR1*C.7)+(0.3*((X(I)-Y2)**2))
      VAR2=(VAR2*C.7)+(0.3*VAR1)
```

Program for the analysis of the filters with "correct delay" (end)

```
C DERDE-ORDER FILTER(THIRD ORDER FILTER)
      T1=C.5*T1+C.5*X(I)
      T2=C.5*T2+0.5*T1
      T3=C.5*T3+C.5*T2
      DEV1=C.5*DEV1+C.5*(X(I)-T3)**2
      DEV2=C.5*DEV2+C.5*DEV1
      DEV3=C.5*DEV3+C.5*DEV2
      I=I+1
      IF(I.LE.12) GCTC 3C
      V1=V1+(XMEAN-Z)**2
      V2=V2+(XMEAN-Y2)**2
      V3=V3+(XMEAN-T3)**2
      V=SCRT(V)
      SVAR=SCRT(VAR)
      SVAR2=SCRT(VAR2)
      SDEV3=SCRT(DEV3)
      D1=D1+(V-SVAR)**2
      D2=C2+(V-SVAR2)**2
      D3=D3+(V-SDEV3)**2
C OUTPUT PER HOUR
      WRITE(6,4C)TIME,XMEAN,V,Z,SVAR,Y2,SVAR2,T3,SDEV3
   4C FORMAT(T8,A1C,T2C,F6.3,T29,F5.3,T38,F6.3,T49,F5.3,T58,
     +F5.3,T79,F6.3,T91,F5.3)
      J=J+1
      IF(J.LE.95) GCTC 5
C OUTPUT CHECKING OF ALGORITHMES
      V1=SCRT(V1/96.)
      V2=SCRT(V2/96.)
      V3=SCRT(V3/96.)
      D1=SCRT(D1/96.)
      D2=SCRT(D2/96.)
      D3=SCRT(D3/96.)
      WRITE(6,5C)V1,V2,V3,D1,D2,D3
5C    FORMAT(1H1,T1C,'CHECKING OF ALGORITHMES'//
     +T1C,'V1=',F6.3/
     +T1C,'V2=',F6.3/
     +T1C,'V3=',F6.3/
     +T1C,'D1=',F6.3/
     +T1C,'D2=',F6.3/
     +T1C,'D3=',F6.3///)
      STCP
      END
```

APPENDIX 2

Program for the analysis of the filters with "correct gain"

(CORRECT GAIN FILTERS)

```
C INITIALISATION
      REAL*8 TIME
      DIMENSION X(12)
      J=0
      Z=8.6
      VAR=0.17
      Y1=8.6
      Y2=8.6
      VAR1=0.17
      VAR2=0.17
      T1=8.6
      T2=8.6
      T3=8.6
      DEV1=0.17
      DEV2=0.17
      DEV3=0.17
      V1=0.
      V2=0.
      V3=0.
      D1=0.
      D2=0.
      D3=0.
      WRITE(6,15)
15    FORMAT(1H1,T10,'TIME',T20,'XMEAN',T30,'V',T40,'Z',T50,'VAR',
     +T60,'Y2',T70,'VAR2',T83,'T3',T90,'DEV3'///)
C INPUT CARDS
5     READ(5,10)((X(K),K=1,12),TIME)
10    FORMAT(12F5.2,10X,A8)
C CALCULATION OF CLASSICAL MEAN AND VARIANCE PER HOUR
      XMEAN=0.
      V=0.
      DO 20 K=1,12
   20 XMEAN=XMEAN+X(K)
      XMEAN=XMEAN/12.
      DO 25 K=1,12
   25 V=V+(XMEAN-X(K))**2
      V=V/12.
C CALCULATION (ALGORITHMES) FOR MEANS(Z AND Y2)
C AND VARIANCES(VAR AND VAR2) PER HOUR
      I=1
C EERSTE-ORDER FILTER(FIRST ORDER FILTER)
   30 Z=(Z*(11./12.))+(X(I)*(1./12.))
      VAR=(VAR*(11./12.))+(((X(I)-Z)**2)*(1./12.))
C TWEEDE-ORDER FILTER(SECOND ORDER FILTER)
      Y1=(Y1*(1.-(1./SQRT(12.))))+(X(I)*(1./SQRT(12.)))
      Y2=(Y2*(1.-(1./SQRT(12.))))+(Y1*(1./SQRT(12.)))
      VAR1=(VAR1*(1.-(1./SQRT(12.))))+(((X(I)-Y2)**2)*(1./SQRT(12.)))
      VAR2=(VAR2*(1.-(1./SQRT(12.))))+(VAR1*(1./SQRT(12.)))
```

Program for the analysis of the filters with "correct gain" (end)

```
C DERDE-ORDER FILTER(THIRD ORDER FILTER)
      T1=(T1*(1.-(1./(12.**(1./3.)))))+(X(I)*(1./(12.**(1./3.))))
      T2=(T2*(1.-(1./(12.**(1./3.)))))+(T1*(1./(12.**(1./3.))))
      T3=(T3*(1.-(1./(12.**(1./3.)))))+(T2*(1./(12.**(1./3.))))
      CEV1=(CEV1*(1.-(1./(12.**(1./3.)))))+(((X(I)-T3)**2)*(1./(12.**(1
     +/3.))))
      CEV2=(CEV2*(1.-(1./(12.**(1./3.)))))+(CEV1*(1./(12.**(1./3.))))
      CEV3=(CEV3*(1.-(1./(12.**(1./3.)))))+(CEV2*(1./(12.**(1./3.))))
      I=I+1
      IF(I.LE.12) GOTO 30
      V1=V1+(XMEAN-Z)**2
      V2=V2+(XMEAN-Y2)**2
      V3=V3+(XMEAN-T3)**2
      V=SQRT(V)
      SVAR=SQRT(VAR)
      SVAR2=SQRT(VAR2)
      SCEV3=SQRT(CEV3)
      C1=C1+(V-SVAR)**2
      C2=C2+(V-SVAR2)**2
      C3=C3+(V-SCEV3)**2
C OUTPUT PER FCUR
      WRITE(6,40)TIME,XMEAN,V,Z,SVAR,Y2,SVAR2,T3,SCEV3
   40 FORMAT(T8,A10,T20,F6.3,T29,F5.3,T38,F6.3,T49,F5.3,T58,F6.3,T70,
     +F5.3,T79,F6.3,T91,F5.3)
      J=J+1
      IF(J.LE.95) GOTO 5
C OUTPUT CHECKING OF ALGORITHMES
      V1=SQRT(V1/96.)
      V2=SQRT(V2/96.)
      V3=SQRT(V3/96.)
      C1=SQRT(C1/96.)
      C2=SQRT(C2/96.)
      C3=SQRT(C3/96.)
      WRITE(6,50)V1,V2,V3,C1,C2,C3
50    FORMAT(1H1,T10,'CHECKING OF ALGORITHMES'//
     +T10,'V1=',F6.3/
     +T10,'V2=',F6.3/
     +T10,'V3=',F6.3/
     +T10,'C1=',F6.3/
     +T10,'C2=',F6.3/
     +T10,'C3=',F6.3///)
      STOP
```

DISCUSSION

VAN ULDEN

Do you think Dr.Cocquyt, that the Gaussian model does apply in extreme circumstances? My suggestion is that you cannot use this model to calculate the 95%-value because it does not apply to circumstances resulting in the highest values.

COCQUYT

Computation of 95% values was done on the basis of counting the values. With the meteorological data, pollution was computed and at each point distribution functions were calculated and these are independent of whether the model is Gaussian or not.

BULTYNCK

The 95%-values were calculated for isolated stacks. If we had considered the weather conditions connected with these 95%-values, we would possibly have found a high wind speed or a high instability and low wind speed. And for these circumstances the model is applicable.

GREENFIELD

Dr.Cocquyt, if I understood correctly you are using the Gaussian model for a distance of 22 kilometers with one wind at the point. Have you attempted to validate the calculations in any way? One thing we know about Gaussian models is that they do not apply a few kilometers from the source, particularly when you use them with one wind.

COCQUYT

We are going to validate these data. The experimental data that we have, indicated that far from the stack the model does not apply very well.

GREENFIELD

I would maintain the model is not good any more than a few kilometers from the stack. I think that a lot of experience that we have run into is that to try and use a simplified model like the Gaussian for long distances and one wind is just an operation of utility.

ZWERVER

The lecture given here was mainly a comparison of two different computing methods, which are very important for authorities in their decision making.

COCQUYT

We did it in order to have at least a basis for making decisions.

BULTYNCK

We have seen that bi-Gaussian models are working for isolated stacks with a ratio between calculated value and the observed value ranging from 1 to 3 and that it is working reasonably well up to distances of 10 to 15 kilometers. This is valid for non-active chemical materials. The model Dr.Cocquyt proposed here has been used for calculating annual averages in the Antwerp region using 200 stacks. The calculated distribution over the town corresponds quite well with the observed pattern.

GREENFIELD

Does the mentioned ratio apply to a single stack?

BULTYNCK

It was for a single stack. It was a result of tracer experiments carried out at our nuclear research centre. We have checked the mathematical model by pouring into the atmosphere fluorescein at the same height and at the same location that a vector-anemometer was situated [1].

GREENFIELD

How many times did you do these experiments and how many points did you measure?

BULTYNCK

We have only carried out 16 experiments in one place and 10 experiments in another. We had some 80 sample locations for each experiment in the down wind direction. What was quite interesting in those experiments was on one hand the good correlation between the y's at different distances calculated from the actual measurements and the y's derived from the vectorial anemometer measurements, and on the other hand that the distance at which the maximum concentration was found, which is an expression of the vertical distribution, also corresponded quite good.

NORDØ

We computed it for Oslo and saw that about two thirds of the transport of air pollutants was due to this thermal circulation. With low wind speeds completely clean air was found downstream of Oslo while a maximum of pollution occurred in the centre of the city. This was due to the inflow of air from the surroundings towards the heating source.

This example shows that one must be very careful when applying the plume formula even close to the source.

COCQUYT

In fact the big problem is to calculate the plume rise. We used the Stümke formula but this was only done to see what results it gives. I am not defending any model.

DEUBER

During the excursion to the Dutch measuring network a value of about 160 $\mu g/m^3$ was detected and we asked what action would be taken but it was shown that at the moment nothing would happen because there are no threshold values. As far as I know the U.K. also handles the problem without establishing limits and I would like to ask Mr.Stewart how the problem of transforming measuring results into action is handled in the U.K.

SCHNEIDER

To understand that nothing was happening this morning, you have to know the situation for which control action is envisaged. When the Clean Air Act came into force in 1970 it was decided that action could only be enforced by the provincial authorities. We have 11 provinces in Holland and the highest authorities in the province have got the power to act as far as the Clean Air Act is concerned. Of course they are not going to do this on their own and they are establishing a service within each province which advises them on the ways in which they can act.

You can think of control or action or traffic distribution, you can think of intermittent control on industry, you can think of the control on power plants, you can think of the supply of fuel for domestic heating. All these measures are being spelled out at the moment. There is the danger that each province will do it in its own way, and according to the Clear Air Act they could. To follow up on our side, as the monitoring network, our responsibility is to advise each of the provinces when they should act, but the decision whether they are going to act is left to them. So the only thing we can do here in Bilthoven is to come to an agreement with each province when and at what level or at what combination of levels or in what situation they would like to be alerted. And they can do this in different steps of course.

But as you can hear from my story, the action is left decentralized. And there is a good reason for this as the different provinces have different problems. We have a few provinces which have real problems with petrochemical industries, there are also provinces which have very specialized industries and the ways of controlling these industries can be quite different. So I agree this is not a direct answer as to what will be done in future, but the way we envisage it is that as soon as the whole network is operational we hope to have an operators' guidebook.So that they know that to do for each province. And we as an Institute hope that all the measures taken in each province will be more or less the same.

STEWART

The control in the U.K. is really anticipatory rather than retrospective in that our Clean Air Act encourages the change of domestic fuel used particularly to eliminate the problem of smoke from that source and to introduce the use of smokeless fuels. So the actual measurements that we make are in effect, monitoring the effectiveness of this. In some regions in the U.K. the introduction of smokeless zones is behind other regions and of course the measurements there bring some pressure to bear on the local authorities concerned to take the necessary action because the action is, as has been indicated in Holland, taken at a local level. We have also some special circumstances. Central London is one area where the levels of SO_2 are higher than is thought acceptable and there is a power taken there to introduce low sulfur fuel to be used in that area on the requirement of the authority. Our basicphilosophy towards this was to anticipate the likely problem and to take action at the source and have presumptive limits on the level or the concentration of emission that is permissable or allowable.

The monitoring activity is to ensure that such enforcement at the source does result in a quality of air which is in alignment with acceptable criteria. I would feel that this system is working satisfactory in that the introduction of this has resulted in a reduction of smoke of about 60 percent in the U.K. over 10 years and a reduction of the ground level concentrations of SO_2 of about 30 percent. Our feeling is probably that we make too many measurements rather than too few but because pollution is a local problem, and there is a local interest in it, it is very difficult to provide a statistical indication that fewer measurements are sufficient because individuals in certain areas will say well; yes statistically no doubt this is satisfactory but we like to know what the levels are in our area. And so for that reason the coverage of the 1200 sites that I indicated continues in the U.K. because of the local interest of the cooperating local authorities in having these data. The Warren Spring laboratory being in that context a coordinating body to publish the results.

REFERENCE

1 H.BULTYNCK, L.MALET, *Tellus* 24 (1972) 5

RELATING AIR QUALITY DATA TO STANDARDS

L.J.BRASSER

Research Institite for Public Health Engineering TNO, Delft, The Netherlands

SUMMARY

The position in The Netherlands concerning air quality standards is described. A compilation of the proposed standard for sulphur dioxide is given.

As air quality data can be presented as - mostly lognormal - frequency distributions, the standard has to agree with this phenomenon. Some ways of comparing standards with measured air quality situations are presented, among others using a testindex.

This last way for the presentation of air quality proves to very versatile.

INTRODUCTION

Conducting air quality measurements is not an objective in itself, the measurements have to serve a purpose. When starting measurements, this purpose has to be described first as it greatly determines the details of the study that has to be made.

Site selection, sampling duration and frequency and data processing are strongly related to an objective of the study.

It is not possible to describe all possible reasons for which an air pollution measuring system can be installed, so here only a number of the most frequent reasons will be given as:

- compiling inventories of the air pollution situation
- study of the influence of air pollution on the health of a population
- study of the influence of air pollution on vegetation, animals,materials
- air pollution warning
- research, for instance study of air pollution phenomenology
- relating the situation to air quality standards

It is clear that in many cases a study can serve more than one purpose.

Whatever this purpose may be, the data produced by the study have to be digested in some way. Storage of the data in a "data graveyard" has no

TABLE I

Standard (24-hourly means, $\mu g/m^3$)

	50 percentile	98 percentile
sulphur dioxide	75	250
OECD standard smoke	$<$ 30	$<$ 90

Standard for transition period (24-hourly menas, $\mu g\ SO_2/m^3$)

	50 percentile	98 percentile
industrialized and/or urbanized areas		
low smoke area	120	350
high smoke area	90	275
areas that now have high pollution concentrations		
low smoke area	150	350
high smoke area	125	275

Definition of the high and low smoke area (24-hourly means, μg OECD standard smoke/m^3)

	50 percentile	98 percentile
low smoke area	$<$ 30	$<$ 90
high smoke area	$>$ 40	$>$120

When the standard smoke concentration is between that of the two defined areas, the SO_2-concentrations in the standard are also selected between their two areas.

sense.

In this paper only the way air quality data are processed in The Netherlands to compare them with air quality standards is presented.

NATIONAL STANDARDS IN THE NETHERLANDS

At this moment there are no national air quality standards in The Netherlands. Several years ago a committee of the Health Council was asked by the Minister of Health and Social Affairs to prepare a proposal for a standard for sulphur dioxide. As a result of this a proposal was made [1]. The committee was asked to proceed with their work (for the new ministry of Health and Environmental Hygiene) and to produce proposals for measuring techniques, for data handling and for air quality standards for a number of other pollutants. At this moment the committee is busy fulfilling this task.

This paper concentrates on the question of data handling but as the proposals for this part of the work of the committee is not ready yet, the paper cannot be seen as the work of the committee but only as the contribution of one of its members.

THE PROPOSED STANDARD FOR SULPHUR DIOXIDE

This proposal is not yet incorporated into the Air Pollution Act but since its publication it has been used on a growing scale. It is based on the lognormal frequency distribution for sulphur dioxide as it is found under most conditions. The selected maximum concentrations for sulphur dioxide are connected to the 50- and the 98- percentile of the frequency distribution of the concentration. As there is an interaction between sulphur dioxide and smoke, the proposal also deals with smoke concentrations, expressed as standard smoke. The proposal consists of two parts, one dealing with the situation that should be reached in a number of years and another less stringent part acting as a standard during a transition period.
Details of the standard are given in table I

THE FREQUENCY DISTRIBUTION OF AIR POLLUTION CONCENTRATIONS IN RELATION TO AIR QUALITY STANDARDS

It is well known that in most cases (not all) the frequency distribution

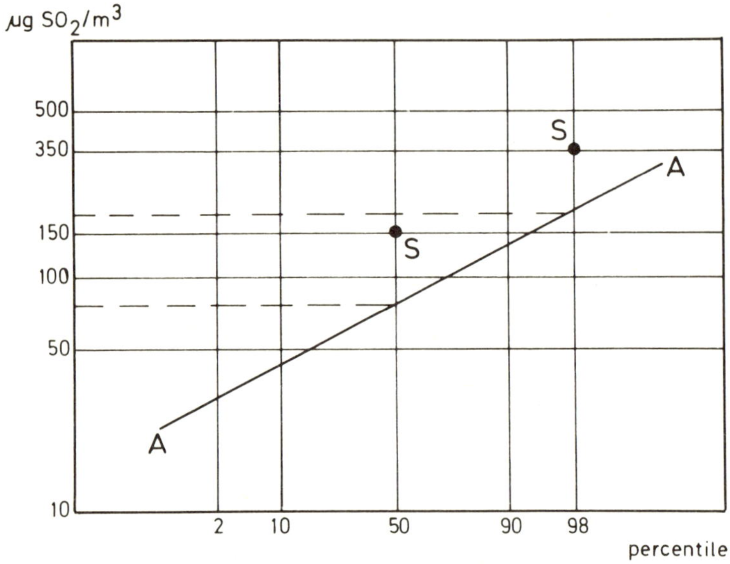

Fig. 1 - Percentile versus sulphur dioxide concentration (pollution lower than standard)

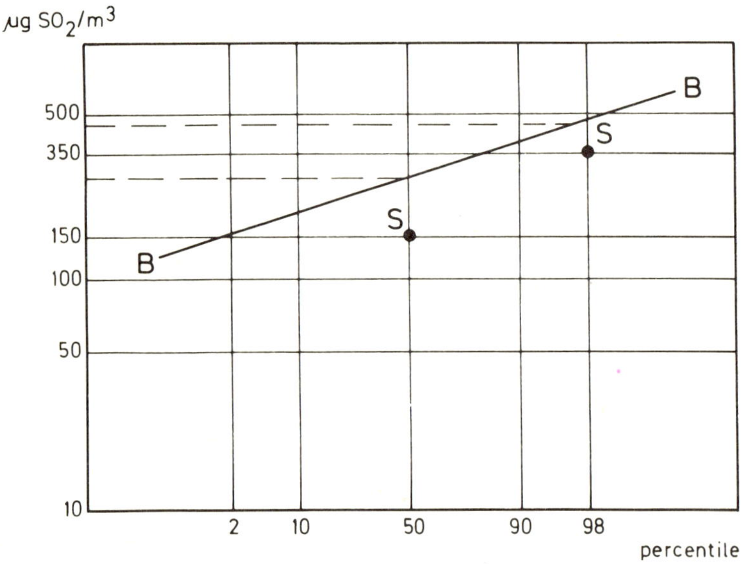

Fig. 2 - Percentile versus sulphur dioxide concentration (pollution higher than standard)

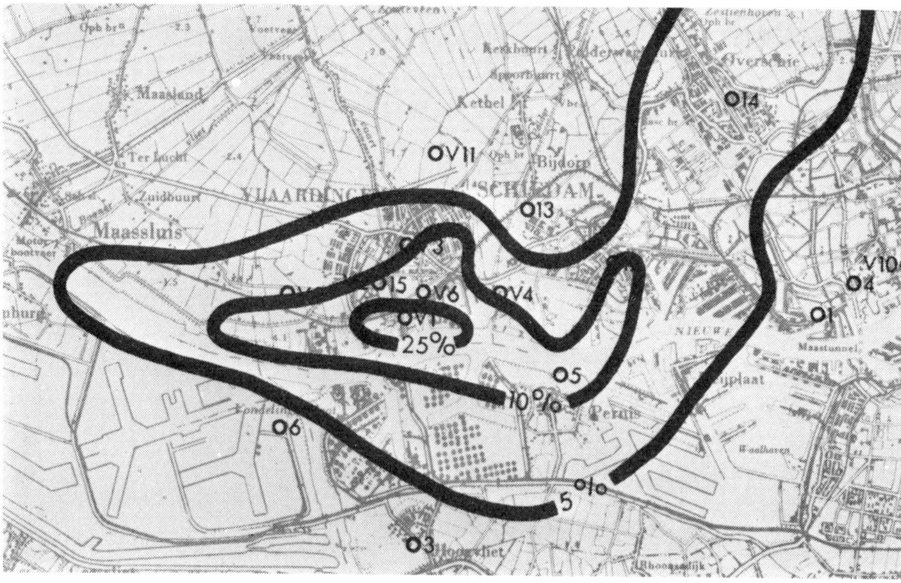

Fig. 3 - Iso-lines of equal excess of a standard concentration

of air pollution concentrations fits or nearly fits the lognormal frequency distribution. This means that it is not possible to set as a standard a maximum concentration that should never be exceeded [2]. In practice every concentration lower than *emission* concentrations will be exceeded sooner or later. So in a standard for *immission* concentrations the maximum concentration should be combined with a percentage of time that the concentration may be exceeded. In The Netherlands this is achieved by setting maximum concentrations for sulphur dioxide for the 50-percentile (the median) and the 98-percentile of the frequency distribution.

As peak concentrations are higher for shorter measuring times than for longer ones, the measuring period has also to be selected and mentioned in the standard. For sulphur-dioxide a measuring period of 24 hours has been selected in The Netherlands, thus producing frequency distributions of 365 daily measurements per year. For other pollutants it may of course be advisable to select other sampling periods.

When measurements serve more than one objective, it is necesssary to select the shortest sampling period necessary for one of the objectives and to calculate means over longer periods for the other objectives.

So for instance in the national air pollution monitoring network in The Netherlands, which is in fact continuously recording, normally one-hourly means are calculated but for the purpose of comparing the data to the proposed sulphur-dioxide standard the data are reduced to 24-hourly means.

The presentation of frequency distributions as graphs or tables may be good for persons accustomed to this type of presentation, it is not workable as a general presentation. Other ways of presentation and/or calculation give more information to the specialists but are still difficult to understand for the general public [3], [4] ,[5].

A method of presenting air quality data in such a way that the data can be mapped without losing too much information (as occurs when yearly or periodical means are mapped) was developed in The Netherlands [6].

This method is also based on the frequency distribution of air pollution concentrations.Starting from a year of measurements in a network of monitoring stations, the frequency distributions of concentrations for all stations are determined. Now in each distribution, the percentile of the measurements that exceeds a certain concentration is read. In practice the concentrations mentioned in a standard can be taken. The percentiles that are found in this way can be mapped, producing lines of equal excess of a given concentration (Fig. 3). As the standard gives an allowed excess of that concentration it can be seen at a glance on the map in which part of the country the standard is met and in which part this is not the case.

Though the lines give some information on the measure of excess, this information is still difficult to read for non-specialists. For this reason another step was made, as described in the next paragraph.

CALCULATING A TEST-INDEX

In a new study [7] another method for comparing a measured situation at a monitoring station with a standard was developed.

The foregoing paragraphs have described how a standard is set in this country. It has also been pointed out that the frequency distribution is determined. From this distribution the cumulative distribution curve is constructed in paper with a lognormal division. In most cases this produces a straight or a nearly straight line.

In the following scheme, this diagram is drawn schematically and the standard is also presented in the diagram as two points (the 50-percentile and the 98-percentile). Two diagrams give two situations, one for a station where the pollution is lower than the air quality standard, a second for a station where the pollution is higher than the standard.

Now the test index T_i can be calculated. The test index is defined as the quotient of the 50- and 98-percentile concentration of the frequency distribution divided by the standard concentrations at the same percentiles. It will be clear that the test index is 1.0 when the situation exactly meets the standard. It is 1.0 when the standard is not met and it is 1.0 when the situation is better than the standard. The test index is a clear measure how far the situation exceeds or is under the standard for the percentiles of the standards.

The test index is calculated as

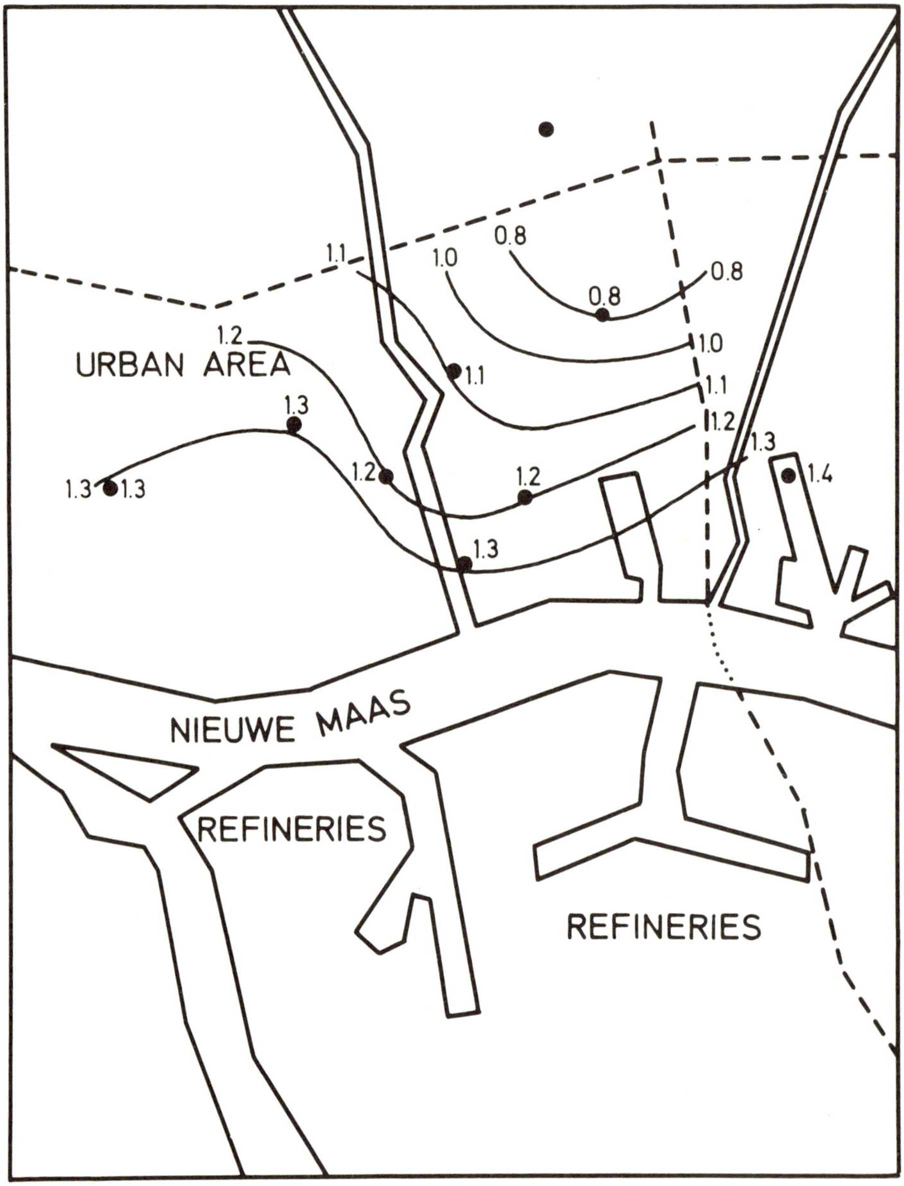

Fig. 4 - Iso-lines of T_{i98} values (city of Vlaardingen-Rijnmond area)

$$T_{i\ 50} = \frac{conc_{50}}{S_{50}} \text{ and } T_{i\ 98} = \frac{conc_{98}}{S_{98}}$$

In the case of the schemes this is for line AA (meeting the standard):

$$T_{i50} = \frac{75ug/m^3}{150ug/m^3} = 0.5$$

$$T_{i98} = \frac{180ug/m^3}{350ug/m^3} = 0.5$$

For line BB (not meeting the standard):

$$T_{i50} = \frac{270ug/m^3}{150ug/m^3} = 1.8$$

$$T_{i98} = \frac{460ug/m^3}{350ug/m^3} = 1.3$$

In practice it is also possible that only one of the values of T_i's is over the standard. Then the situation does not meet the standard. The values of T_i then give information whether the situation is too high because of high median concentrations ($T_{i\ 50} > 1.0$) or because of high peak concentrations $T_{i\ 98} > 1.0$).

Mapping values of T_i for the different measuring stations gives information on where the standard is exceeded (Fig.4).

VERSATILITY OF THE DATA HANDLING SYSTEM USING A TEST-INDEX

The system as described in the foregoing paragraph was originally developed for air pollutants having a log-normal distribution of frequencies of concentrations. In fact the system can be used in connection with an air quality standard stating one or more concentrations with their acceptable limits of excess.

This means that it can be used for every air pollutant showing a continuous frequency distribution for which a standard of the type described has been selected. Though the maps should be drawn by specialist, reading them does not give difficulties to non-technical people.

REFERENCES

1. Advies inzake grenswaarden SO_2. *Verslagen en mededelingen van het Ministerie van Volksgezondheid en Milieuhygiëne*, Vol.1971, no.22

2. L.J.BRASSER, P.E.JOOSTING and D.VAN ZUILEN, Sulphur dioxide - To what level is it acceptable? *Report G300 of the Research Institute for Public Health Engineering TNO, Delft, The Netherlands, July 1967*

3. R.I.LARSEN, Determining reduced emission goals needed to achieve air quality goals - a hypothetical case, *Journal of the APCA,* December 1967, Vol.17, no.2

4. R.I.LARSEN, A new mathematical model of air pollution concentration, averaging time and frequency, *Journal of the APCA,* January 1969, Vol.19, no.1

5. R.I.LARSEN, Relating air pollutant effects to concentration and control, *Journal of the APCA,* April 1970, Vol.20. no.4

6. L.J.BRASSER, A new method for the presentation of a large number of data obtained from air pollution survey networks, *Paper presented at the Second International Clean Air Congress,* December 1970, Washington D.C.

7. L.J.BRASSER, Presentatie van meetuitkomsten van luchtverontreinigingsconcentraties in verband met normen voor de beperking daarvan, *De Ingenieur,* 7 January 1972, Vol.84, no.1

EVALUATION OF MEASURED SO_2 CONCENTRATION BY TRANSFORMING TO A PROBABILITY VALUE

S.KRUIZINGA

Royal Netherlands Meteorological Institute, De Bilt, The Netherlands

INTRODUCTION

An automated SO_2 measuring network can serve many purposes. Two of these are connected with a short-term warning system or with research. In these cases one wishes to have a data handling system which can help to answer the following questions.

a. Which measured SO_2 values indicate that the atmospheric dispersion is so poor that it is necessary to limit the discharge of bad smelling gaseous wastes.

b. Which situations are worth further meteorological analysis.

For both these questions, an indicator for atmospheric dispersion would be useful.

It is our aim to develop such an indicator based on SO_2 measurements. We will assume that hourly measurements of SO_2 concentrations are made. The indicator is based on the following considerations. The factors that influence the SO_2 concentration can be divided into non-meteorological factors and meteorological factors. Now the whole set of data of a sampling station can be divided into groups of data which are measured in roughly the same non-meteorological circumstances. The basic idea is that the highest SO_2 concentrations in such a group are coupled with bad meteorological conditions especially when other stations give relativily high concentrations at the same time. Indication of interesting hours can simply be done by denoting the fraction of hours in the same group that gave lower SO_2 concentrations. This fraction can also be interpreted as the probability of measuring a lower concentration than the actual one in the same non-meteorological circumstances. So our proposal is to transform the SO_2 concentrations to some type of Probability Measure (to be called P.M.).

CLASSIFICATION OF THE NON-METEOROLOGICAL FACTORS

As stated in the introduction, we wish to compare the SO_2 concentration

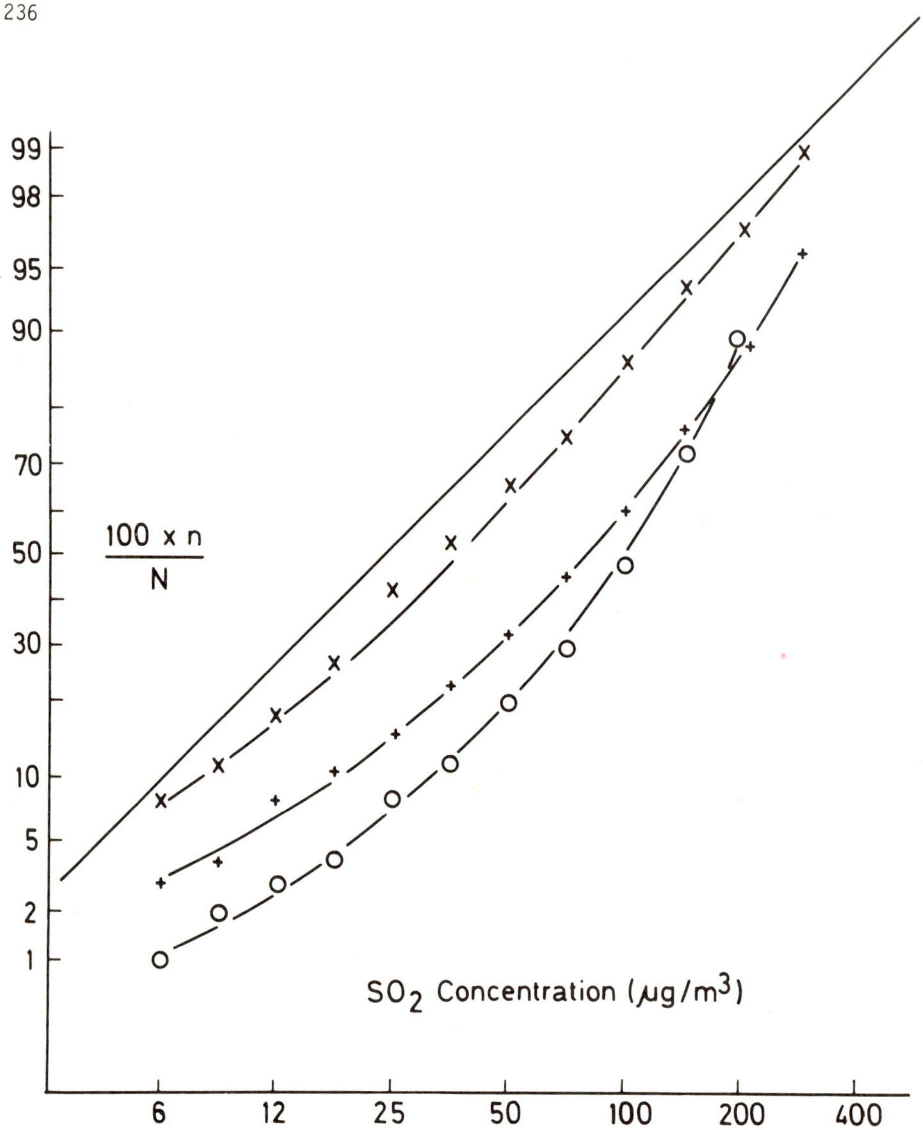

Fig. 1 - Cumulative frequency distributions of hourly values in comparison with the log-normal distribution

Station 19

o windsector 8 N = 159

+ windsector 1 N = 296

x windsector 2 N = 254

at a certain hour with the concentrations measured at other hours with roughly the same non-meteorological circumstances. So it is necessary to divide our material into classes on the basis of the non-meteorological factors. In our pilot study we have taken only two non-meteorological factors into account:

a. The emission. This factor is roughly constant on short term but has a definite yearly cycle caused by the seasonal changes in domestic heating.
b. The wind direction. This factor is partly of meteorological importance but most of its influence is due to the geometrical patterns of detectors and sources.

The class to which an hourly SO_2 concentration belongs is determined by the time of the year and the wind direction. Therefore the windrose has been divided into eight sectors of 45 degrees each, and a ninth group is reserved for variable winds.

The year on the other hand has been divided into twelve periods of thirty days, the measurements of these periods will be compared with the measurements from the associated ninety day periods.

TRANSFORMATION TO THE PROBABILITY SCALE

The simplest way to transform a specified SO_2 concentration to a P.M. is by counting the hourly values in the given class that are lower than the given value and dividing this number (n) by the total number (N) of hourly values in that class. On a computer this is a simple job which requires, however, many operations and a large memory. A more advanced solution is by fitting a theoretical frequency distribution to the hourly values in a given class and computing the P.M. with the help of the parameters of the frequency distribution. For SO_2 concentrations the log-normal distribution is widely accepted. In our case, however, it turned out that this distribution was not always usable as is shown by Fig. 1 where some experimental distributions are given. In order not to be dependent on the assumption of log-normality we have used the following procedure.

At a number of boundary values g(k) the probability of finding a lower concentration was estimated with help of the following expression.

$$P\ (\mu_{SO_2} = g\ (k)\ ) = 100 \times \frac{n}{N}\ \text{(percent)}$$

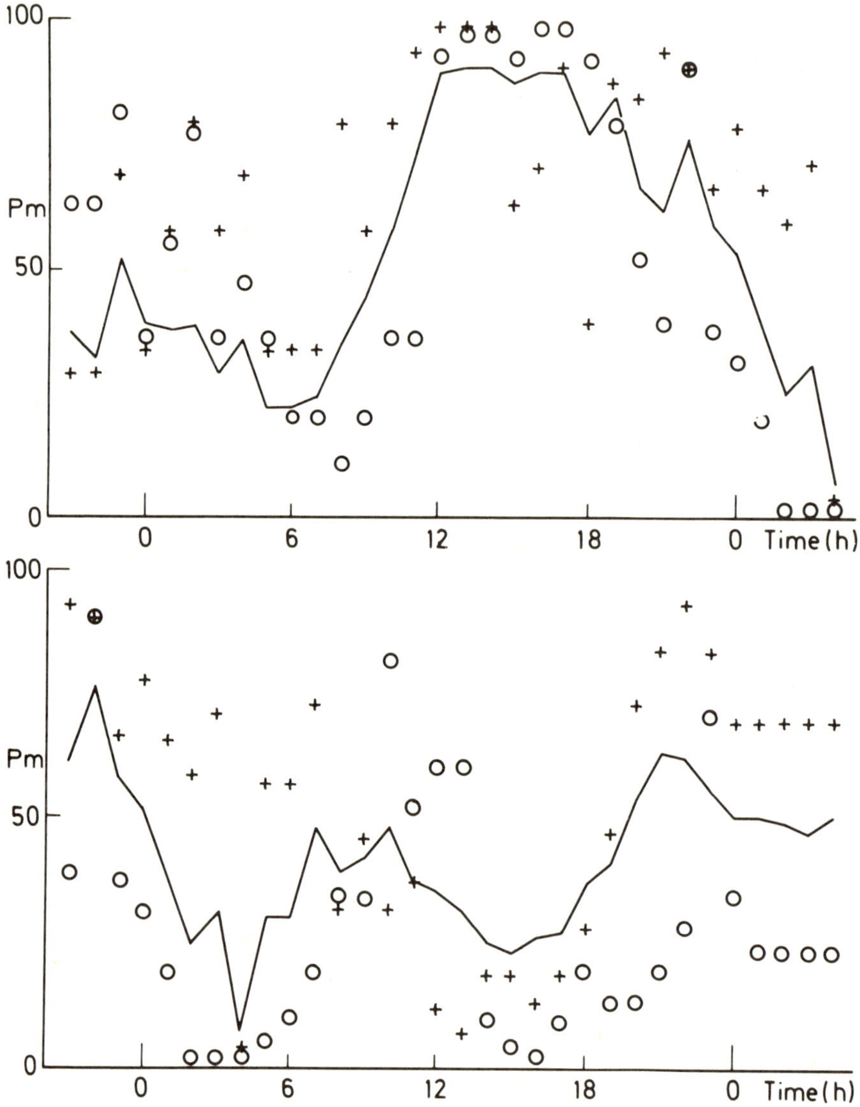

Fig. 2 - P.M.-values on 3 and 4 October 1971 plotted versus time, for two stations of the Rijnmond Network (5 and 19), and the area mean P.M.-value

+ station 5

0 station 19

- area mean

Thereafter the probability measure P related to a specific SO_2 concentration μ_{SO_2} was determined as follows:

If $g(k-1) < \mu_{SO_2} < g(k)$

then $P = \frac{P(g(k-1)) + P(g(k))}{2}$ percent

EXPERIMENTAL RESULTS

The system described has been tested with the help of SO_2 measurements from the network of the Openbaar Lichaam Rijnmond. This network is situated in the harbour area of Rotterdam-Schiedam and consists of 31 stations which automatically deliver hourly the mean SO_2 concentration to the central computer.

We have analysed two thirty-day periods, the first running from 27-8-1971 up to 26-9-71, the second from 26-9-1971 up to 26-10-1971.

The reference period consisted of the thirty day period itself, the preceding thirty days and the following thirty days. Some of the results are shown here. In Fig. 2 we have plotted the P.M. as a function of time for two stations as well as the mean of the P.M. of all stations. In this figure two days are given. The first day is a known bad day in which many complaints about bad smells were received by the authorities. On the next day the number of complaints was much lower. As is seen, this coincides with the fact that on the first day the mean P.M. of the stations becomes 90% during a couple of hours while at the second day the mean P.M. stays below 65. A P.M. of 90% must be interpreted in the following way. For ninety percent of all the hours in the reference period with the same wind direction, a lower concentration was measured at that station than at the given hour and of course at ten percent a higher concentration was measured.

According to our proposal this high P.M. indicated that the dispersion of the atmosphere is relatively poor. If this is true it is necessary that all stations have high P.M. independent of the actual SO_2 concentration.

In Fig. 3 the spatial distribution of both the P.M. and the SO_2 concentration is depicted. This figure gives a situation with high P.M. Circles denote the sampling stations, the station numbers are placed within the circles. The SO_2 concentration and the P.M. are given above and below to the right of the circle respectively. We see that although there is a considerable variation in SO_2 concentrations, the P.M. is rather high at most of the stations. Stations 28 and 29 especially are showing a great contrast in actual

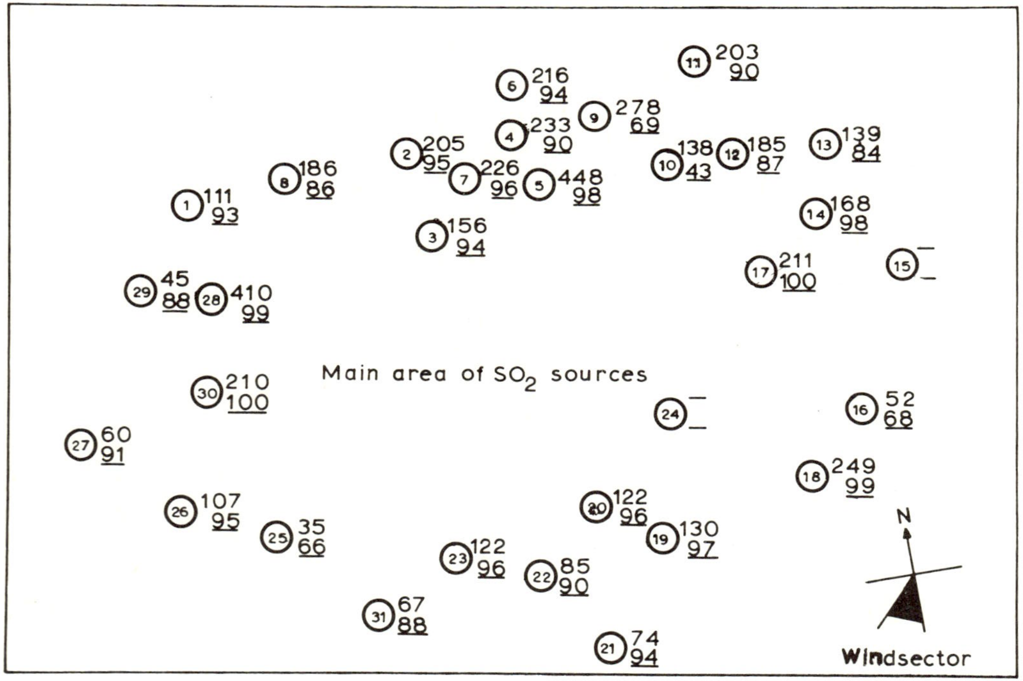

Fig. 3 - Spatial distribution of SO_2 concentrations and P.M.-values on 3 Oct. 1971 - 14.00 hr.

Station number $\frac{SO_2 \text{ conc.}}{\text{P.M.-value}}$

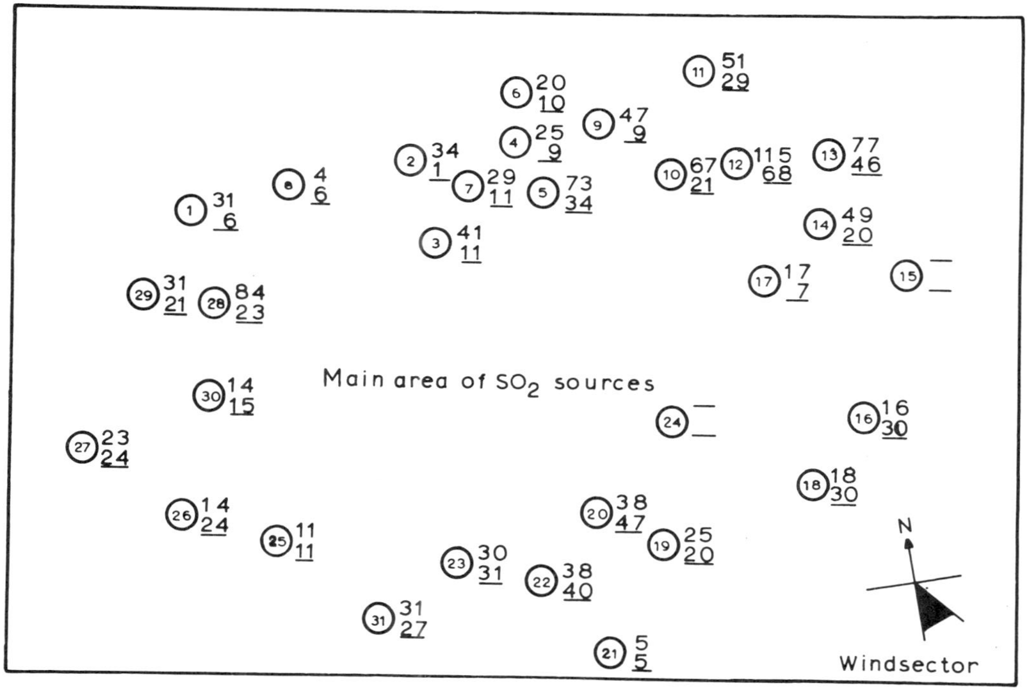

Fig. 4 - Spatial distribution of SO_2 concentrations and P.M.-values on 3 Oct. 1971 - 6.00 hr.

Station number $\frac{SO_2 \text{ conc.}}{\text{P.M.-value}}$

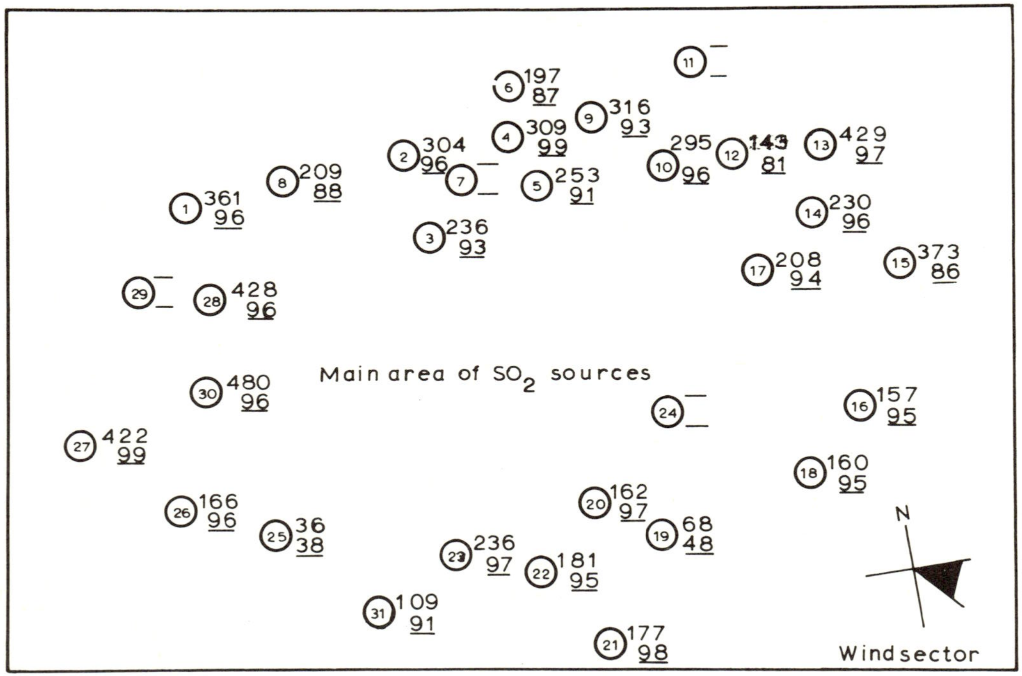

Fig. 5 - Spatial distribution of SO_2 concentrations and P.M.-values on 30 Oct. 1972 - 11.00 hr

Station number $\frac{SO_2 \text{ conc.}}{\text{P.M.-value}}$

SO_2 concentrations. The greatest distance depicted in this figure is about fifteen kilometers.

In Fig. 4 a situation with low P.M. is depicted. It has the same main features as the situation shown in Fig. 3. But it is interesting to note that concentrations higher than the lowest of Fig. 3 are nevertheless indicated as quite normal. This is caused mainly by the fact that these concentrations are measured at other stations which obviously have higher normal concentrations.

By searching for high mean P.M. it is possible to select interesting situations for meteorological studies. For the warning system it is not possible to wait until all measurements needed for the reference distributions are performed. For this case we propose to use the reference distributions, prepared for a thirty-day period of this year, in the same period of next year. When the reference distributions are ready it is possible to perform the transformation to the P.M. on-line on the central computer. In Fig. 5 a typical high P.M. situation in 1972 is depicted. The transformation was performed on the basis of measurements for 1971. The situation has the same main features as the situation shown in Fig. 3. In spite of the considerable variation in SO_2 concentration nearly all the stations have a high P.M. and thus indicate that the dispersion is poor.

The indicator described has also been compared with a purely meteorological indicator, This meteorological indicator has been developed by Dr.Van Ulden of the Royal Netherlands Meteorological Institute and was computed with help of the following meteorological data:

1. Difference between the potential temperature at 500 meter height and the minimum temperature at 1,5 m.
2. The daily mean of the wind velocity.
3. The maximum mixing height.

Its scale runs from zero to ten corresponding respectively with very good dispersion and almost no dispersion. It gives, however, only an indication for the mean daytime dispersion. Therefore we have compared it with the daytime mean of the area mean of the P.M. In Fig. 5 these two indicators are plotted for a number of days. As is seen the two indicators are fairly well correlated. The correlation coefficient is about 0.70 but the relation need not to be linear.

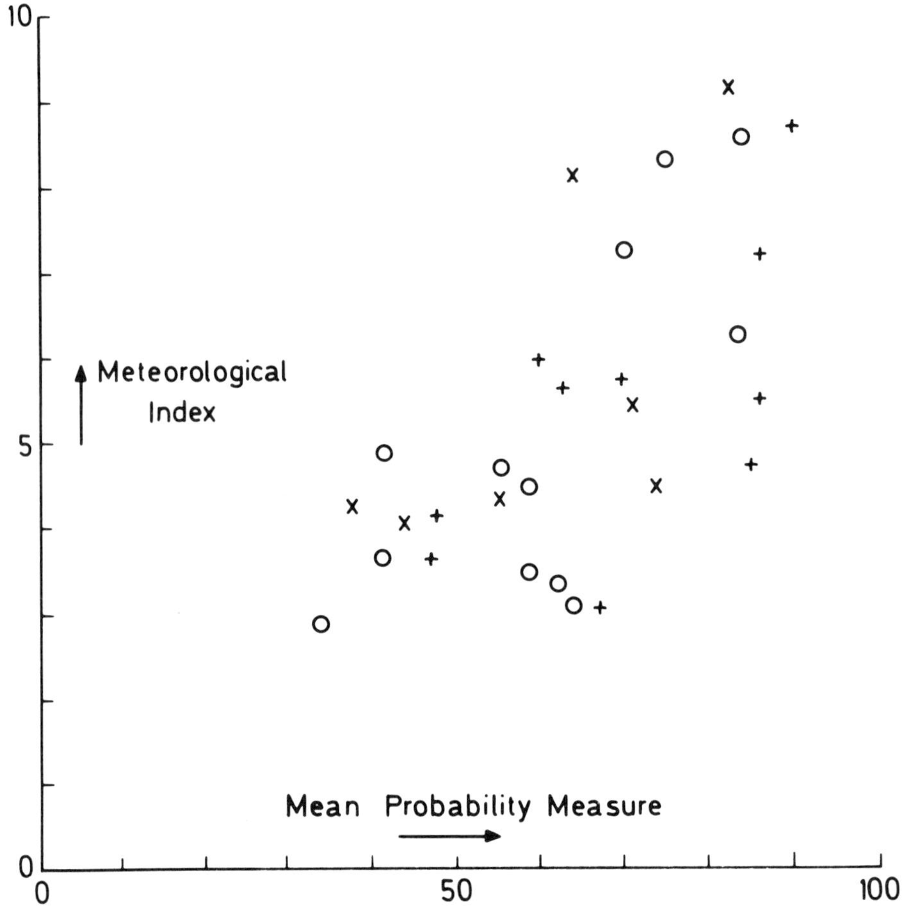

Fig. 6 - Meteorological index plotted versus the mean P.M.-values for a number of days with respect to different reference periods

+ 16 to 26 Sept. 1971	Ref.Period 1 in 1971
0 26 Sept. to 80 Oct. 1971	Ref.Period 2 in 1971
x October 1972	Ref.Period 2 in 1971

CONCLUSION

By transforming the measured SO_2 concentrations of all stations to the probability of finding a lower concentration at that station one gets a quantity, called Probability Measure, which shows a simple spatial relation. Moreover, it is concluded that this P.M. shows a reasonable correlation with a Meteorological Index for the dispersions of the atmosphere. So it is possible to use this P.M. as an indicator for atmospheric dispersion. It is proposed to use this P.M. for the indication of interesting situations for further meteorological analysis. With some slight modifications it is also possible to use this P.M. for a short-term on-line processed warning system.

The ideas described here resulted from discussions in a commission for the countrywide SO_2 network of the Netherlands.

DISCUSSION

JANTUNEN

What should be done when the standard is exceeded?

BRASSER

When the standard is set, it is the task of the Minister to tell what should be done with it. In this paper only the view of the Committee is expressed, not the view of the Minister himself. The committee has decided that this standard at it is here should not be exceeded by measurements in any station. But I do not think that the Minister will act vigourously when one station in a certain area just exceeds this standard. But when a number of stations in the same area are over the standard, as you have seen on one of the maps, then there is a reason for acting.

In the Westland area this has been done by asking for the use of very clean natural gas instead of the high sulphur content oil.

VAN ULDEN

The 98 percentile may cause some trouble. It depends rather strongly on meteorological conditions which vary from year to year. So it is difficult to set the standard on the 98 percentile of one year.

BRASSER

I completely agree with you and the committee was aware about that. It is the reason that the 50 percentile is one of the major percentiles to look at. But of course we do not like to have very high peak concentrations. There have been very long talks on the appropriate percentile. My personal view is that 98 is still high in the frequency distribution so that perhaps 95 or 90 would have been better.

STEWART

There is some danger, when using this percentage, of losing sight of the fact that pollution is episodic. Dr.Brasser touched on this. It is likely that very often the important periods will be where the levels have persisted. Have you any suggestions how these percentages might be weighted? It is not important that we have 10 minutes each day over the year but perhaps that there are some days days in winter which account for almost all the annual "two percent".

BRASSER

It is the reason that the committee went to the 24 hourly measurement in the first place, to take away high peak concentrations of very short duration.

The 98 percentile is seven days per year. So a five day period nearly consumes all the available days of high concentration. That is another reason why it would be better to go a little bit lower.

P.SPAANDER

Dr.Brasser mentioned two methods of data presentation. Which method is preferable for the future?

BRASSER

There are two methods that have been used and are being used at the moment. The first one has been used in a very wide sense in the western part of the country and all scientists are very happy with it. The reason for the test index was that the meaning of the percentages was not quite understood by politicians. So for the scientists the percentage method is preferred, because it gives more information. For the government, the members of parliament etc., the test index is preferred.

SHEARER

Considering annual averages some limits must be put on the number of times per year that the maximum value can be exceeded. The way in which our U.S. standards are set up allows an annual average for sulphur dioxide of 80 $\mu g/m^3$ with a maximum 24h concentration of 365 $\mu g/m^3$, not be exceeded more than once per year. If one is not careful one can get oneself tied to close to an annual average, meanwhile exceeding a number of times the maximum 24h concentration.

ZWERVER

We have experience with smog periods and we advise pollution abatement during episodes. The system we use is the same as is described by Dr.Kruizinga. Is it a new approach?

Dr.Kruizinga mentioned the words "warning system" and "prediction". In my view it is not a prediction or warning system because the situation at that moment is described.

KRUIZINGA

I do not really know if this method is quite new. Prediction is mentioned only in the sense of meteorological prediction and I stated that for the meteorological prediction, preliminary meteorological research was needed. This research can best be done in periods that are interesting with respect to high air pollution levels.

For a short term warning system one is too late if the P_m values are high. We do not have a short term warning system on a meteorological basis, only a global prediction system exists. It is just a description of the actual situation and it does not predict bad situations in the future.

VAN ULDEN

I would like to give a further answer on the prediction possibilities of this method. A sharp increase in the index, shortly after sunrise, indicates that you may expect difficulties in the Rijnmond area. We found out that a sudden rise of this index in the morning almost always corresponds with a high number of complaints and with high concentrations of other pollutants. In other words, a sudden rise means there is fumigation and if a meteorological forecast is available, there is a prediction possibility for periods of a couple of hours.

STEWART

It seems to me that if you are using the data from the previous year that you are getting quite good success, in fact 98 percent for most of the stations. One wonders therefore if the time might be envisaged when most of the stations will become unnecessary because you have been able to predict what could happen at that point with 98 percent probability of success.

KRUIZINGA

If you drop out stations you can only state something about the probability that a certain value would be measured. The actual concentration is not really known. If the reference distribution is known it should be possible to compute the actual concentration from the mean P_m at each site.

BERLIN

Is it not more simple just to say that a value should not be exceeded over seven days rather than to talk about the 2 percentile? Further, do you feel, as you already expressed, that taking the whole year period is more reasonable than taking for example the winter period? In some areas, for example Northern Scotland, summer values were higher than the winter values while in other areas there is a ratio of seven between the winter and summer averages.

BRASSER

The percentiles can be translated into a number of days and that is what is done in the reports. But for calculating the test index a percentile is needed.

Some comments on the second part of your question. If all SO_2 comes from domestic heating then all high concentrations will occur in winter. In the Westland area the main source is the heating of greenhouses. Heating is started not when the days are getting cold but at the moment that the days are getting longer. The heating period is extended from the beginning of January up to May. So the period of high concentrations is later and therefore it is better to use one year for every possibility. Then you cover everything. But when you are only looking for one simple type of source, you can use all the winter values, like we do for some towns, and have a very good impression: in that case the high values are all during the winter.

This is however, not normal. We intend to use this way of measurement also for other pollutants where circumstances may be quite different.

GOOTJES

In the paper of Dr.Kruizinga wind is mentioned as a meteorological factor. Is it the most important one or are there others?

VAN ULDEN

In the meteorological index and in the forecast a potential temperature difference is included which is very strongly related to cloud cover. Wind is only a very small factor.

ANALYSIS OF SIX YEARS CONTINUOUS AIR POLLUTION SURVEILLANCE

D.JOST, R.KALLER, H.MARKUSCH, W.RUDOLF

Pilot Station Meteorological Institute, Frankfurt University, Germany

Low wind velocity (annual average 1.8 - 2.7 m/sec) and frequent temperature inversions characterise the air pollution conditions in the industrial area of Frankfurt/Main between the mountains of Odenwald, Spessart and Taunus. Due to the topographic situation, the prevailing winds are from SW and NE [1].

Population, traffic density industrialisation and the resulting emission of air pollutants were increasing in this region when an automated air pollution surveillance system was started in Frankfurt. The Meteorological Institute had already conducted some research on air pollution distribution both by continuous monitoring and by mobile observations from automobiles and aircraft [2, 3, 4]. From these measurements the highest SO_2 concentrations were to be expected in the centre of the city with secondary maxima in the western and eastern part of Frankfurt [4]. According to these findings, the surveillance was started with 3 stations, one central station (CO_2, CO, SO_2, C_nH_m, NO + NO_2, particulate matter, wind velocity, wind direction, temperature, relative humidity, radiation balance) and two satellite stations (SO_2, CO), which were completed by another 5 sites (Fig.3). The technical organisation of the station can be seen in Fig.1; Fig.2 shows how the data handling procedure is set up. The system enables continuous check and correction of the measuring instruments to be automated by the control computer and by external inputs. The routine data output delivers ½ h averages and 1 d averages of all parameters as well as variances of some selected parameters. These data are typed and punched on paper tape and are evaluated by a larger computer. From these mean monthly and daily variations, frequency distributions and cases where concentrations are in excess of set levels are derived [6, 7].

The comparison of air pollution concentration data with air quality standards leads immediately to the problem of the significance of measuring sites with respect to the measuring time and the location of these sites. The necessary significance of air pollution data also depends on the requirements which are to be met by the surveillance system.

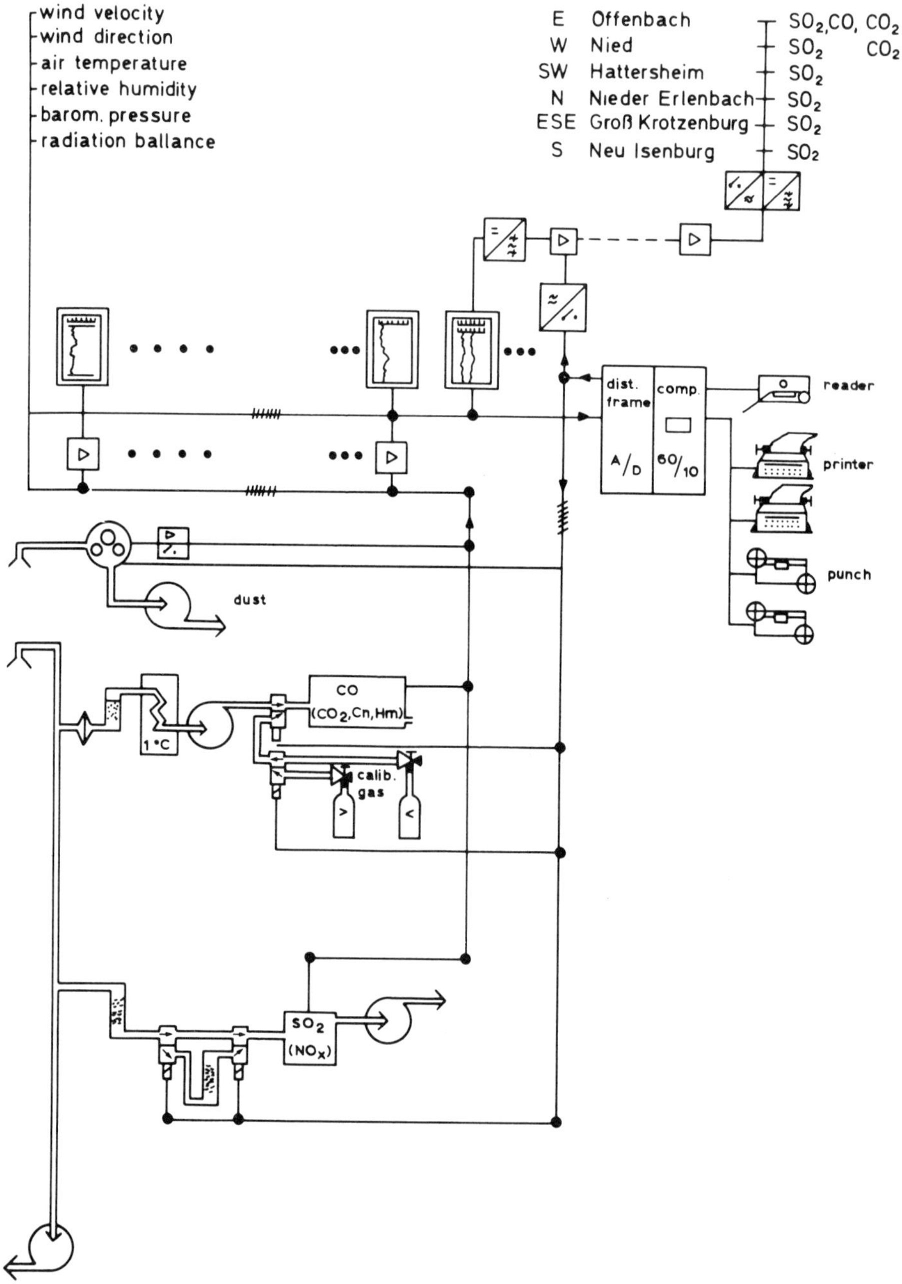

Fig. 1 - Pilot station Frankfurt/Main

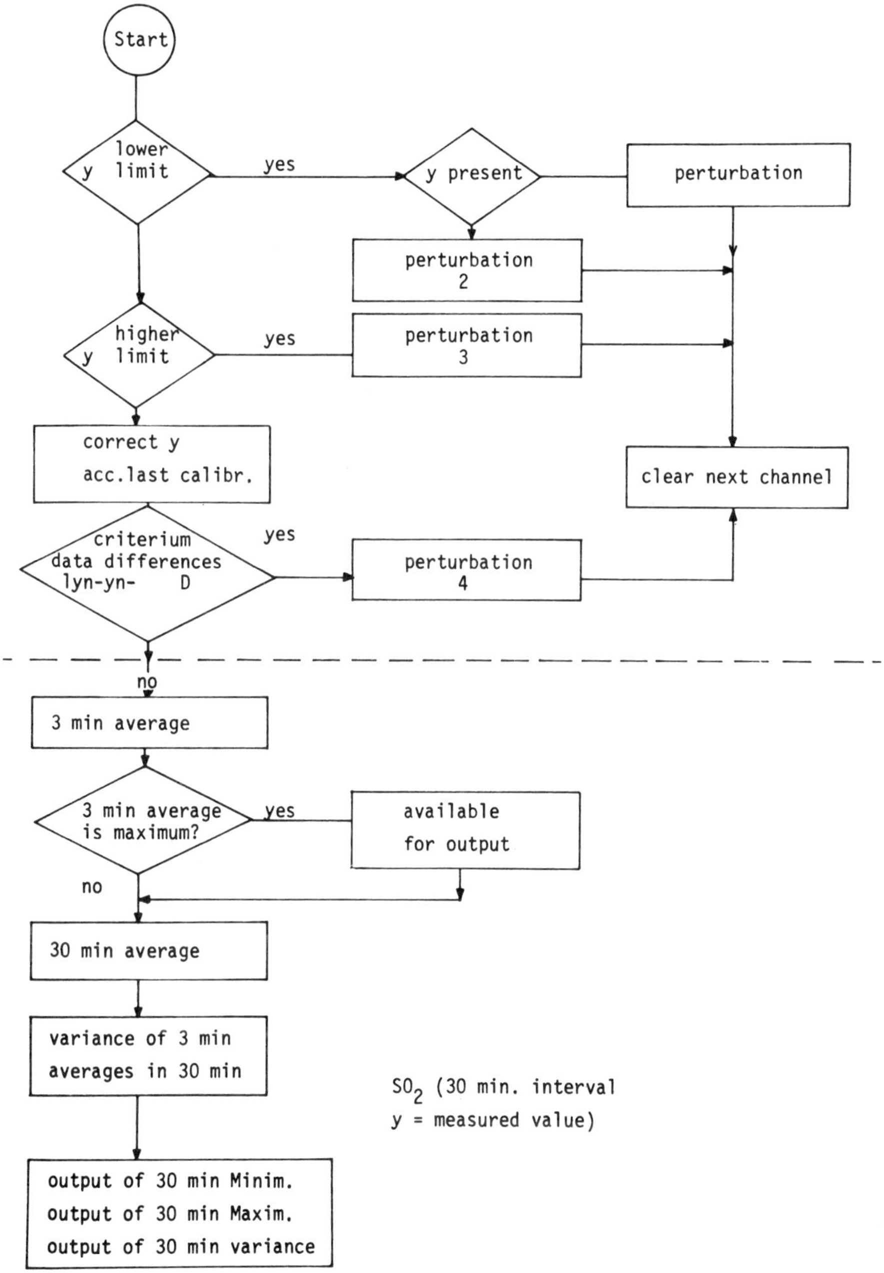

Fig. 2 - Data reception and data handling

Table I - Air pollution surveillance requirements and how to meet them

Surveillance requirements / Tasks of surveillance	discontinuous density of sites		continuous density of sites				
	incidental	permanent	incidental	permanent	on line	off line	data to be evaluated
Episodes with high conc.				$3/100km^2$ (8,9) a	X		actual values
Trend of concentration		$6/100km^2$ b		$3/100km^2$ (8) b,c		X	average and percentiles
Regional check of threshold values		$100/100km^2$ d		$6/100km^2$ a, b		X	average and percentiles
Single source-oriented check of threshold values	$100\text{-}400/km^2$ e	X e	$6\text{-}12/100km^2$ e	X e	X f	X	average and percentiles met.data important
Multiple source-oriented check of threshold values	$100/100km^2$ e	X	$6/100km^2$ e	$6/100km^2$ e,f (11)	X f(11)	X	average and percentiles met.data important actual values, f
Data for city and regional planning	X	$100/100km^2$ d	X	$6/100km^2$ c		X	average and percentiles met.data important
Data for epidemiological studies			X	X	X	X	average and percentiles including met.data
Basic research	all modes	possible	depending	on kind	of research		

a. depending on source distribution and on components which are to be checked. Example for SO_2

b. depending on development of source distribution

c. incidental mode with mobile stations possible, e.g. sites changing yearly

d. incidental mode with one measurement per two weeks

e. depending on size of source

f. only this mode permits influence on emission

g. mode of operation depending on kind of studies

X this mode is possible

Different requirements of air pollution surveillance system resulting from case studies are given in Table 1.

This table is based mainly on results from SO_2 monitoring. The problems of traffic exhaust surveillance are noted. Measurements of oxidants are tested now and indicate anthropogeneous O_3 concentrations during special weather conditions [14]. The number of 6 sites/100 km^2 in Table I is taken from observations with 1 site per 4 x 4 km. The numbers indicated represent minimum requirements for the Frankfurt/Main area based on six years air-pollution surveillance experience. The numbers depend strongly on source distribution, topographical and meteorological situation in the area to be monitored. Studies on traffic exhaust immission surveillance are still continuing. Economic aspects of operating an air pollution surveillance system are not considered. Hickey et al [12] demonstrated that an on-line operation is less expensive than other modes when 24 or more instruments (6 sites with 4 measuring instruments each) are integrated into the system. These considerations lead to an automated surveillance system with about one station per 4 x 4 km. Some of these may be operated with annual change of location and supported by incidental discontinuous measurements (Fig.3). Stations in prevailing wind sectors, in the most densely populated area and around industrial sources were chosen in Frankfurt.

Experience from air pollution studies in Frankfurt shows strong correlations between some kinds of pollutants, especially during high concentration episodes. Total particulate matter correlates with SO_2 and NO_x, Pb with CO, but SO_2 does not correlate well with CO. This means that during air pollution episodes SO_2 and CO may be used as reliable tracers. Measurements of total hydrocarbons yielded no correlation to other pollutants, nor to the whole pollution situation. This result is supported by Dutch studies which show that hydrocarbons should be differentiated [13] . But whereas for SO_2, 3 sites per 100 km^2 seem efficient enough, CO surveillance requires more stations in traffic areas. The location of traffic-oriented measuring sites should be proved by theoretical dispersion studies in order to avoid the necessity for CO-measuring instruments in all streets with heavy traffic.

During air pollution episodes as well as during the remaining time, the concentrations of air pollutants show regular variations distinguished by source variation (for instance SO_2 from heating, CO from traffic), and by variation of meteorological conditions which is reported elsewhere [8, 2, 3, 15]. An example indicating regular variations of SO_2 concentration and indicating the effect of measuring time on the cumulative frequency distribution is to be seen in Fig.4. For 10 min. sample duration the highest values

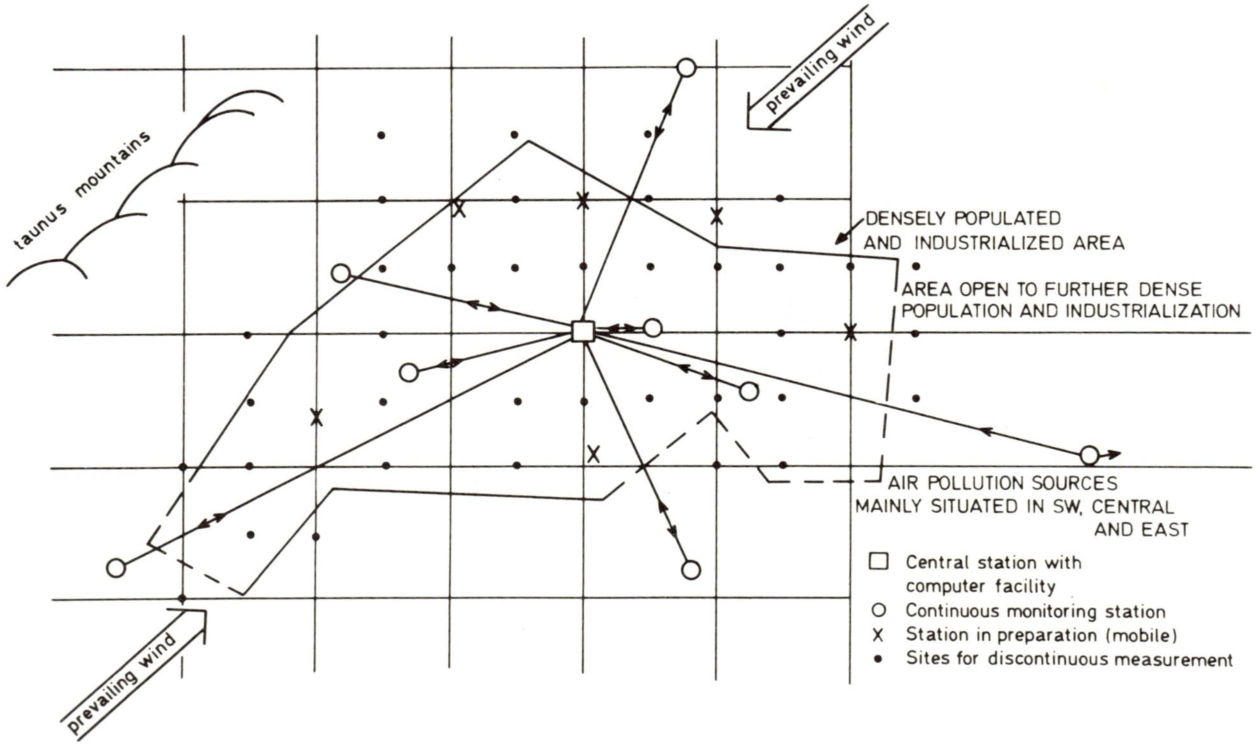

Fig. 3 - Air pollution surveillance

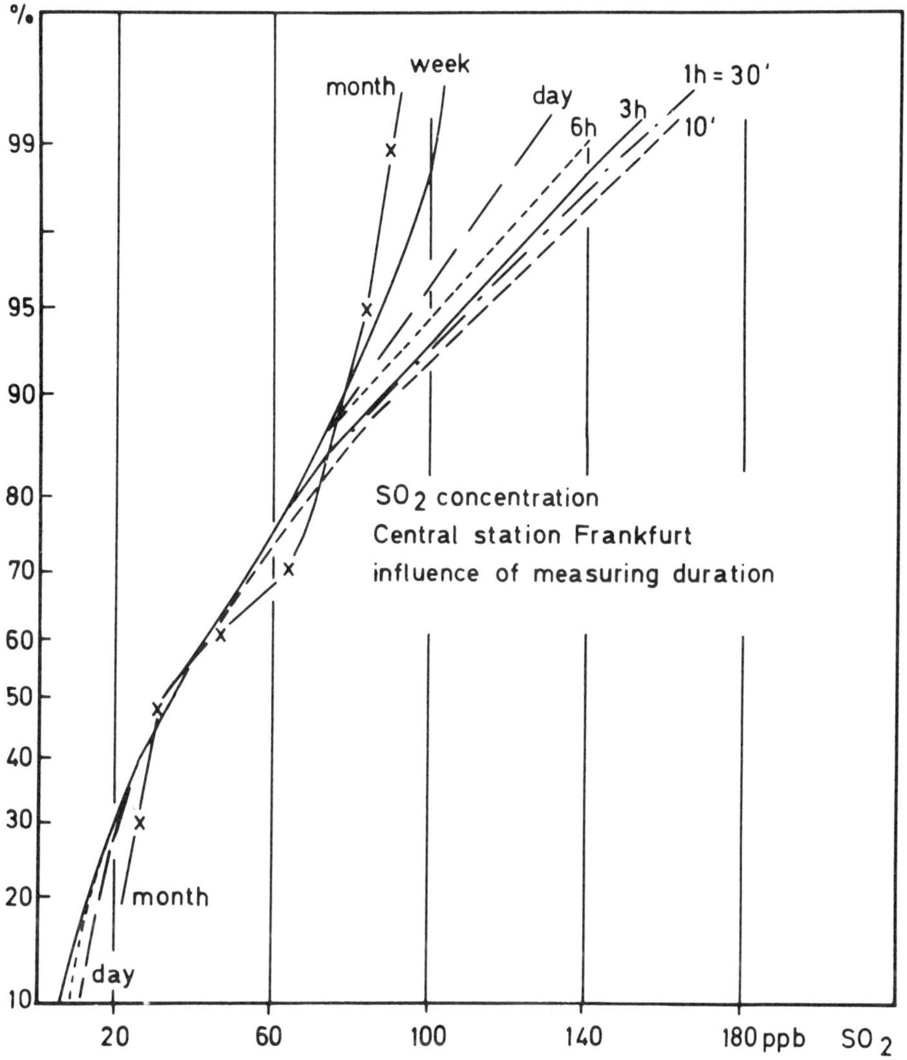

Fig. 4 - Effect of measuring time on the cumulative frequency distribution

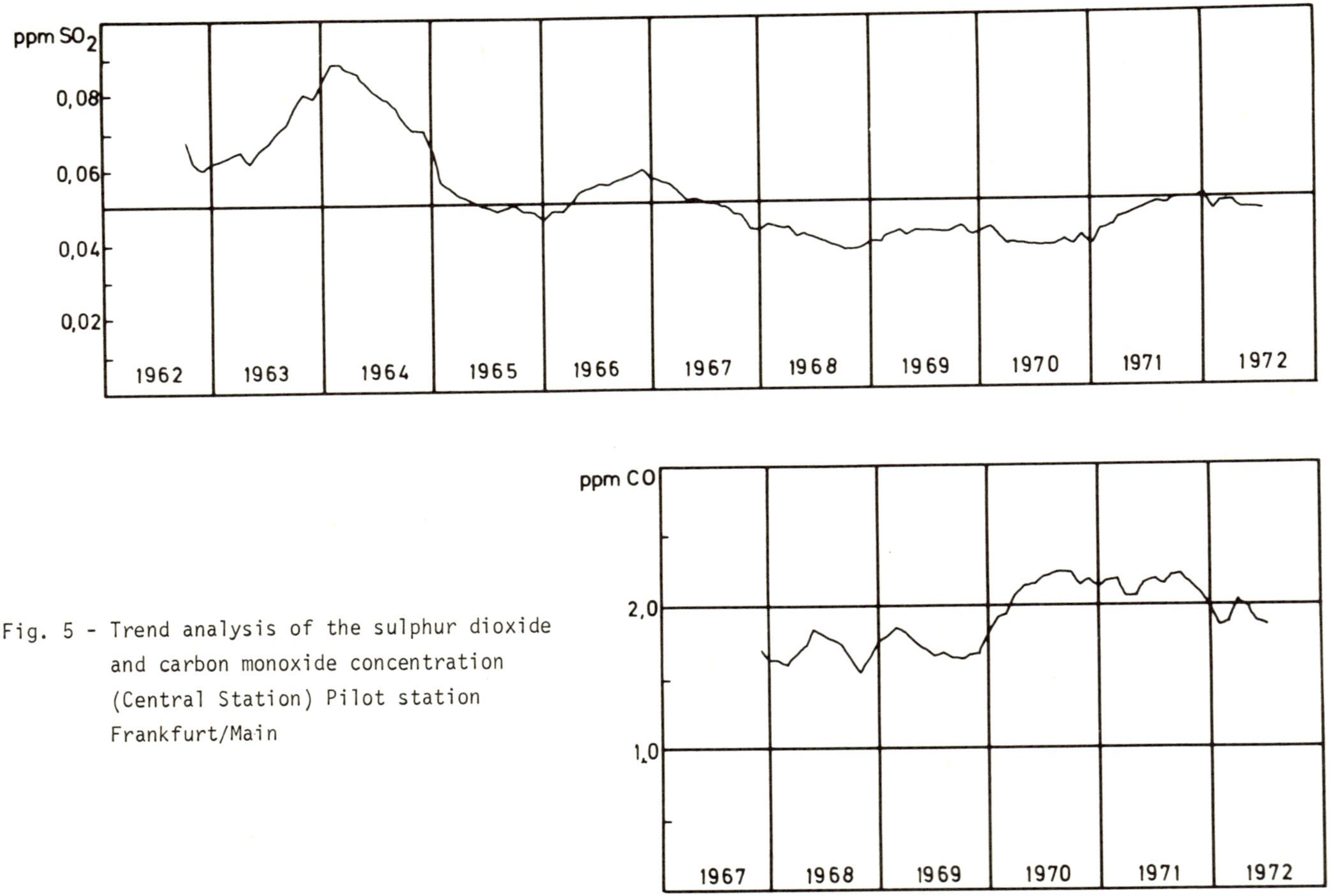

Fig. 5 - Trend analysis of the sulphur dioxide and carbon monoxide concentration (Central Station) Pilot station Frankfurt/Main

are received beginning with the 60 percentile. Up to the 90 percentile there is no significant difference between 30 min. and 3 h sample duration. The longer sample durations, from 6 h to 1 month even, yield identical average values but significant differences for the higher percentiles. The relatively large spread of concentration data for 10 min. sample duration may be due to inhomogeneous source distribution around the measuring site.

Beginning with the 6 h periods the influence of regular meteorological variations becomes obvious (daily variation, variations with synoptic scale, yearly variation). The influence of sampling duration is dependent on the periodic concentration changes [16]. These results do not fit values derived from theoretical considerations, which indicate relatively large spreads for sample durations shorter than one day [17]. Comparing successive monthly averages, the long-time trend of air pollutant concentration is as shown in Fig. 5. Fig. 5 gives an example of SO_2 and CO trends in Frankfurt/Main central station. The SO_2 decrease from 1963 to 1965 cannot be due to meteorological variations as can be seen from comparison with wind velocity, wind direction frequency of temperature inversions and temperature itself. Since 1968, both CO and SO_2 concentrations increase, more for CO than for SO_2.

These studies are supported by "Bundesminister des Innern" under "Pilotstation Frankfurt".

REFERENCES

1 H.WACHTER AND H.SCHARRER, *Ber.Inst.Met.Geophys.Univ.Frankfurt*, 22,(1970)

2 H.W.GEORGII, H.DOMMERMUTH AND E.WEBER, *Ber.Inst.Met.Geophys.Univ.Frankfurt*, 14,(1968)

3 H.W.GEORGII, E.BUSCH AND E.WEBER, *Ber.Inst.Met.Geophys.Univ.Frankfurt*, 14,(1967)

4 D.JOST, *Annalen der Meteorologie*, nr.6,(1973),151

5 E.WEBER, *Schriftreihe des Vereins f.Wasser-, Boden- und Lufthygiene*,30, (1970), 19-47

6 E.WEBER,*Lufthygieneische Monatsberichte der Pilotstation Frankfurt* (since 1971)

7 E.WEBER, *Wissenschaftliche Arbeiten zur Luftreinhaltung, Univ.Inst. Met.Geophys. Frankfurt/M. Anual reports*

8 E.WEBER, *Wissenschaftliche Arbeiten zur Luftreinhaltung Univ.Inst.Met. Geophys. Frankfurt/M, Comprehensive report 1972*

9 D.JOST, *Staub*,30 (1970) 296

10 H.J.KORNER, Sulfur-Dioxide and dust settlement measurements with the screen net method in the area of lower Main Region. *Arbeitsberichte zu: Lufthygienisch-meteorologische Modelluntersuchung in der Region Untermain.*

11 L.A.CLARENBURG, Is it possible to conduct a rational air pollution management. *Report 72-8 by Rijnmond Authority, Schiedam (The Netherlands)*

12 L.A.CLARENBURG, Milieubelasting voor Luchtverontreiniging. *Report by Rijnmond Authority, Schiedam (The Netherlands)*

13 H.R.HICKEY, W.D.ROWE, F.SKRIMER, *Journ.Air.Poll.Contr.Ass.* **21 (1971)** 689.

14 R.GUICHERIT, Luftverunreinigung in Verkehrstunneln. *Report by TNO, Delft, The Netherlands, 1972*

15 D.JOST, Survey of the distribution of trace substances in pure and polluted atmospheres. *Pure and Applied Chemistry,* 24 (1970) 643

16 S.BEILKE, Untersuchungen über das Auswaschen atmosphärischer Spurenstoffe durch Niederschläge. *Ber.Inst.Met.Geophys.Univ. Frankfurt/M,* nr.19 (1970)

17 S.KULSKE, B.PRINZ, Untersuchungen über den Einflusz des Messzeitraumes auf Mittelwert und Vertrauensbereich eines SO_2-Immissionswertekollektivs. *Schriftenreihe der Landesanstalt für Immissions- und Bodennutzungsschutz des Landes Nordrhein-Westfalen, Essen,* 12 (1968) 81

18 R.I.LARSSEN, A new mathematical model of air pollutant concentration, averaging time and frequency. *J.Air.Poll.Contr.Ass.* 19 (1969) 416

DISCUSSION

BERLIN

In Fig. 5 of Dr.Jost's paper the trends from 1963 to 1972 are represented and a particular decrease in the winter of 1964 to 1965 is noticed. The paper indicates that this was not due to meteorological conditions and there was no mention made of temperature at all. 1963-1964 was a particularly cold winter, and I wonder of there was no influence from heating. I am surprised that from these data there seem to be no differences in SO_2 levels for Frankfurt in summer and in winter.

JOST

The averaging of the trend analysis is done by overlapping the monthly averages and therefore you will not have a yearly variation. We also tested the influence of temperature on this trend in 1964 on the basis of an empirical equation we got from our Frankfurt data. This test could not prove a decrease of SO_2 concentration.

BRASSER

Dr.Jost mentioned that the data presented in Fig. 4 do not correlate exactly with Larssen's data. What we tried to do in The Hague was to see whether Larssen's arrow head diagram, which is in fact the graphical representation of this mathematical model, would be true. The arrow-head diagrams that we found in the Hague do not entirely confirm Larssen's theory. We find that from the values of one month down to two hours it really fits the theory pretty well, but for shorter durations it does not fit it at all. I agree that it is too theoretical but it can be used in practice.

RUFF

Which applications are you referring to?

BRASSER

As long as you keep to the mathematical model it is very good for scientists but you cannot use it for politicians who have no mathematical or statistical training.That is the reason that we do not try to represent all the data we have in the mathematical formulas, but we make arrow head diagrams. That is still difficult for them to understand.

RUFF

We found that Larssen's model works quite well for most cases in the middle percentile ranges but reaching the extreme values there is some question of how accurate it is and the extreme values are the ones you are interested in. Do you feel that the graphical presentation, the log-normal distribution, is hard for authorities to understand.

BRASSER

These graphs are useful if you want to make a standard. As soon as standards are made, the graphs should be left out.

RUFF

Dr.Jost mentioned that his data do not conform to Larssen's criteria, and this is not unusual, but I would like to know in which respect they did not follow Larssen's criteria.

JOST

The Larssen model does not fit at the extreme values.

RUFF

During a recent symposium it was generally agreed that almost any skewed distribution will fit and not just the log-normal.

SUESS

I would ask Dr.Jost to evaluate the Frankfurt system after six years of operation. Dit this system justify itself? Is it going into routine use now?

JOST

Operating a system with stations four kilometers apart was the most economical mode, compared to the discontinuous measurements we did in Frankfurt. In order to be representative this density is needed. The system will also be used in the future and similar systems are planned in other parts of Germany. For example Nordrhein-Westphalia is starting a system with a similar density. Bavaria has a different strategy since the location of sites is dependent on the position of sources like industrial areas and cities.

SUESS

Many NO_2 premeasurements have been carried out and since little is known about the relation between NO_2 and health effects this would be a good subject for study.

JOST

Epidemiologists are very much interested in NO_2 measurements. Now these pollutants are measured at only three stations; we are asked to include more stations for the measurement of nitrogen oxides, especially for NO.

STEWART

It seems to me, though it will probably be controversial, that it is rather contradictory to take a sample at a single point while we are in fact interested in air pollution in three dimensions. We had very brief glimpse from Dr.Van Egmond of the correlation spectrometer in this context, we also had a very brief glimpse from Dr.Nordø of the use of a satellite to plot the emission from a volcano near Iceland.

I wonder if anyone would like to comment on this subject.

SHEARER

If we continue for an indefinite future to instal large numbers of stations to make up a monitoring system, this will be too expensive.

We have to analyze the cost-benefit aspects of these systems. Are we going to use 10% of the data to make our decisions or are we going to use all data.

We should consider other means by which we can get the same type of information. We have to increase our efforts in the modelling aspects of the prediction of air quality and the relationship of the emission from sources to the air quality.

Emission inventories, coupled with better models could do a pretty good job in predicting what is going on in the environment in terms of air quality. I would agree that we should use more and more sophisticated monitoring devices such as aircraft and satellite monitoring.

We also need to increase our development in the areas of remote sensing. There is an increasing amount of data being turned up for SO_2, particulate matter, etc., which points to the fact that we should be more concerned about such things as nitrates and sulfates. We sometimes tend to forget the synergistic relationship between SO_2 and particulate matter. I want to ask Dr.Schneider, whether measurement of particulate matter at the monitoring stations was considered and I think we may need more research about trace elements, mercury, lead, asbestos, etc. We also should begin to monitor more adequately our background unpolluted areas. Before you can build a complex source such as a shopping centre or a sports complex you must make an evaluation of what is going to be the increase in air pollution. We recently computed a short survey around three of our large shopping centres in North-Carolina and we observed 8 h averages of CO up to 35 ppm.

JANTUNEN

In Finland air quality measurements are carried out in the surroundings of factories and in the cities and our biggest concern is not gaseous pollutants. We are measuring them too, but we are analyzing suspended particulates, using high volume samplers. We are always measuring at least ten trace metals, nitrates, sulfates, fluoride etc. Mostly we are measuring around factories where no other sources of pollution exist. We are also calculating the correlation between the distance to the factory and the amount of trace metals. We would like to know what people are actually breathing, using a very small sampler that could be carried around all the time by certain persons, living in a polluted area.

SCHNEIDER

The reason why we did not measure particulate matter in our network is a very simple one, we have no suitable monitor. We are actually planning to do it in the future. We had the idea that monitoring in such a way that you can discriminate between trace elements would be much important than just monitoring the total amount of particulate matter.

For the future we need long range monitors to fulfil the whole operation. I think there will be some use for point monitoring, because we want to determine what people are actually breathing, and for this the long range instrument is not suitable. We are monitoring a lot of things already and I can see an even longer list before me of things that have to to be done, and actually for the moment I do not see a solution for this problem. Then you can say do not solve it with an automatic system, but when you just think about the amount of work involved, it does not matter whether the system is manual or automatic. At some stages we were called the monitoring fools in Holland, going out into the field and putting some monitor near a cow, trying to monitor the rural air. I think gradually the idea is spreading that you cannot restrict yourself to the real urban monitoring if you want to find the answer to a lot of problems.

CLEARY

The parameters we monitor at this moment are related to the ease of measuring them. When we consider the recommendations of air criteria groups on environmental quality criteria,emphasis should be placed on several pollutants: oxides of nitrogen, manganese,nitrites and nitrates, polychlorinated biphenyls, mycotoxins, asbestos, sulfuric acid, aerosols, fluorocarbons, organo-bromine compounds.

FINAL DISCUSSION

At the end of the general discussion, it was proposed by Schneider that an international study and working group on monitoring systems should be set up. After discussion, the following outline was accepted by all participants at the symposium.

- The members of this international group will be Research Institutes or Government Agencies.
- Experts from the member Institutes and Agencies will be appointed as participants of the study and working groups.
- Among the activities planned are: organization of subsequent symposia (at regular 2-year intervals); preparation of testcases on the planning, construction and use (evaluation) of monitoring systems; publication of relevant data and information in the form of a newsletter; and continuous exchange of technical knowledge concerning concepts in monitoring system.

At the symposium, seven Institutes and Agencies indicated their intention to join such a group. Also, W.H.O. and the European Community expressed their interest in the proposed international co-operation in the field of air pollution monitoring systems. The proposals made are very much in line with the long-term programme of both the W.H.O. Division of Environmental Health and the Regional Office in Copenhagen.

The National Institute of Public Health (R.I.V.) agreed to act as the first Secretariat for a period of 2 years (1973-75). A letter of invitation will be sent to all potential members before the end of 1973. This letter will outline the purposes of the group and will give a suggested list of subjects to be covered in the next symposium, to be held in 1975.